Experimental Methods in Kinetic Studies

Revised Edition

Experimental Methods in Kinetic Studies

Revised Edition

B.W. Wojciechowski
West Bay Club
Estero, FL, U.S.A.

N.M. Rice
Queen's University
Kingston, Ontario, Canada

2003

ELSEVIER
Amsterdam – Boston – London – New York – Oxford – Paris – San Diego
San Francisco – Singapore – Sydney – Tokyo

ELSEVIER SCIENCE B.V.
Sara Burgerhartstraat 25
P.O. Box 211, 1000 AE Amsterdam, The Netherlands

First edition 2001 (isbn 0-9714895-0-5)
Revised edition 2003

Library of Congress Cataloging in Publication Data
A catalog record from the Library of Congress has been applied for.

British Library Cataloguing in Publication Data
A catalogue record from the British Library has been applied for.

ISBN: 0 444 51314 0

♾ The paper used in this publication meets the requirements of ANSI/NISO Z39.48-1992 (Permanence of Paper).
Printed in The Netherlands.

Preface

This text was inspired by the long-standing involvement of one of the authors (BWW) in the pursuit of a fundamental understanding of chemical reaction mechanisms and of the consequent reaction kinetics. At various times he was involved in experimental investigations of the homogeneous gas phase pyrolysis of hydrocarbons, inhibition of gas phase reactions, Fischer Tropsch synthesis, catalytic cracking and catalyst decay. In all of these cases, he and his associates pursued a fundamental understanding of chemical reactions via the formulation of likely reaction mechanisms and the verification of the consequent rate expressions. In this, they differed from most industrial efforts and many academic studies in catalyst research, where rates were measured largely to establish catalyst activity and selectivity in the course of testing of new catalyst formulations. As a result, the kinetics of even well-known catalyst formulations were rarely well established or studied in adequate detail.

It was a concern to us that the study of kinetics as a tool for understanding reaction mechanisms was so uncommon. Personal experience showed BWW that kinetic studies are indeed a tedious undertaking, requiring many runs before sufficient data became available to verify a mechanistic postulate. Yet this is the kind of work that has to be done if we are to unravel the kinetics and mechanisms of reactions. Work of this kind can be of great practical interest in the development of catalysts and of reactors operating in industry.

It is clear that economic considerations and time pressures incline most industrial research workers to settle for standardized catalyst evaluations, using one or a few runs at standard conditions. Those who have access to more resources and the opportunity to investigate catalytic reaction mechanisms are generally inclined to study the details of surface reactions using high vacuum techniques, even though the reaction conditions encountered are far from those applicable in industry. Nonetheless it seems that, by consensus, surface studies acquired the cachet of a "hard science", while mechanistic studies, using kinetics at realistic reaction conditions, continue to languish under a cloud of suspicion regarding their reliability.

The suspicions are not without cause. The number of runs backing up the average kinetic study is usually small, as small as can be tolerated, so as to minimize the time and tedium of experimentation. The data are normally collected over an extended period of time, often months, and in ways that increase the size of random error. This makes rate equation fitting, and consequently the interpretation of this data, less than certain. Finally, because data cleanup procedures are neither well established nor applicable to the small data-sets available from most studies, the experimental data is usually interpreted in its raw form; that is, with little or no attempt made to identify, treat and reduce the experimental error.

Over many years, a few workers have consistently worked to acquire sufficient data to verify mechanistic postulates. This led them to seek increasing levels of reactor automation to relieve the tedium. It has nevertheless become clear with time and experience that no amount of automation of conventional isothermal reactor operation can increase the rate of kinetic data acquisition beyond a factor of something less than ten. This improvement in data production rates is simply insufficient to make the resultant kinetic interpretations significantly more convincing. If the data shortage problem is to be solved, the inadequacies of the existing data collecting methods point to the need for a paradigm shift.

The breakthrough came with the development of temperature scanning (TS) methods by the authors, Professors B.W. Wojciechowski and Norman M. Rice. Unlike the previously known temperature ramping methods, that yield two dimensional accounts of events, temperature scanning collects data that allow the construction of three dimensional reaction surfaces. These contain enough information to allow for the use of sophisticated mathematical procedures for error correction. Moreover, they allow reliable interpolation without using curve fitting. The result is unlimited amounts of unprejudiced data for rate-model evaluation. All that was needed was a method of interpreting the data obtained by temperature scanning so that it would yield conventional rates of reaction, together with the associated temperature and conversion levels.

It turns out that the mathematics behind the required interpretation procedure are straightforward, and require few assumptions. In fact, the mathematical formulation behind temperature scanning could have been written down many decades earlier. One might speculate that it was not formulated sooner because the dominant paradigm of kinetic studies required that rate data be collected at isothermal conditions. But, even if the equations had been written earlier, one would still have had to wait for computers to handle the massive amounts of data manipulation required to arrive at conventional rates of reaction using raw temperature scanning results.

The temperature scanning technique as currently understood is broadly applicable and can be applied to the study of reactions in all phases, as well as on catalysts, as long as certain readily verified requirements are met. The following text is intended as a guide to kinetic studies of reaction mechanisms. It reviews reactor types that may be applicable in various circumstances, data collection methods, methods of error removal, and the interpretation of data collected using temperature scanning reactors. It is our aim to revive kinetic studies as a useful approach to the understanding of reaction mechanisms, and so put catalyst development on a rational foundation. We dedicate this book to that end.

The authors wish to thank Professor Diran Basmadjan for his careful reading of a draft of the first edition of this work, and for his wise and constructive comments. We also acknowledge our debt of gratitude to Margot Wojciechowski, who tirelessly, patiently and cheerfully edited this manuscript to design and produce the following text in camera ready form.

B.W. Wojciechowski
Professor Emeritus, Queen's University,
Naples, Florida, USA

N.M. Rice
Associate Professor, Queen's University,
Kingston, Ontario, Canada

Table of Contents

Introduction

The observed dependence of product selectivities and reaction rates on reaction conditions is completely determined by the underlying chemical reaction mechanism.

Reaction Kinetics - State of the Art and Future Needs

Early studies of chemical reaction mechanisms were intimately connected with efforts to quantify and interpret the pertinent reaction kinetics. The literature of the first half of the twentieth century is full of elegant kinetic mechanisms outlining and quantifying the details of successively more complex reactions. As things turned out, this fruitful interaction between kinetics and mechanistic studies was not sound enough, and the available rate measurement techniques not good enough, to prevent the abandonment of kinetics as a tool in mechanistic studies. As a result, mechanistic studies based on reaction kinetics have lately been neglected in the rush to develop improved industrial processes by means of empirical, or at best quasi-empirical, studies.

The record shows, however, that even in the recent past, during the greatest flowering of empirical R&D in catalysis, mechanistic studies were recognized by a persistent minority to be too important to abandon completely. Despite this, the study of overall kinetic mechanisms came to play a diminished role in catalyst development. Fashion and developments in instrumental methods of analysis combined to attract students of reaction mechanisms to the details of intermediate processes involved in the overall mechanism. During this time, ingenious and expensive equipment for the observation of surface-resident species came to dominate mechanistic studies in catalysis, as in all of chemistry. For example, complex procedures and apparata were devised to study the rates of individual elementary reaction steps in gas phase and surface based mechanisms. But the study of overall reaction rates, and their interpretation using mechanistic rate expressions, foundered on the tedium of the extensive and repetitive experimental work required. This type of study continues to be severely hindered by the need for the massive amounts of data that are required to deal with the difficulties of discriminating between rival mechanistic models. Masses of good kinetic data are also essential to establish the "true" parameters of the model that is finally identified as appropriate for the reaction.

The neglect of kinetics as a science, lasting some decades during the rapid growth in tertiary education, has largely denuded universities of specialists in this field, and left the chemical industry dependent on heuristic rate expressions and a less than satisfactory understanding of reaction mechanisms. Studies of the details of reaction processes, often of individual reaction steps, are done mostly at conditions well removed from those of interest to industry. The knowledge thus gained cannot be assembled to yield either the pertinent overall mechanism or the overall kinetics of industrial processes.

The advent of computers, with their power in dealing with complex calculations, encouraged some to write large arrays of possible intermediate reactions, based on the individual steps observed or supposed, and fit the parameters of these arrays of reactions to conform with overall rate and selectivity observations. Such procedures are of doubtful value, on the one hand because of the lack of well founded information on the elementary rate parameters of individual reaction steps, and on the other by the large uncertainties due to the continuing scarcity of rate data and errors in its measurement. The procedures used in these methods are soundly based but perhaps premature in their application. Not enough is known about the rate parameters of the individual steps to constrain the choices that have to be made in constructing mechanisms in this way to a reasonable extent.

At present these exercises are more interesting than definitive. At the same time, the appearance of the computer has created an opportunity for the application of complex rate expressions in reactor design and in data fitting and parameter estimation. In ways unthinkable before, it has provided us with the means of evaluating the kinetics of complex mechanisms. What computers *per se* cannot do is provide us with the massive amounts of experimental data required for the fitting of complex mechanistic rate expressions. For that a brand new approach to the measurement of reaction rates is required.

It turns out that the data generation problem can be solved by changing the established procedure of operation of experimental reactors. The acquisition of massive amounts of rate data does not depend on the invention of a new reactor configuration, or on the simple automation of the time-consuming procedures used in established reaction rate measurements. A conceptual breakthrough is needed. It is now clear that the concept that had to be challenged was the need to measure rates at steady state and at isothermal conditions. Discarding of this long-standing paradigm is so traumatic that a considerable portion of the following text is devoted to an explanation of the fundamentals of conventional rate measurement, and of the theory and consequences of the proposed fundamental departure from these conventional methods of rate measurement. This new departure in experimentation abandons rate data acquisition under isothermal conditions and instead involves a technique called *temperature scanning*. A reactor using this technique is called a Temperature Scanning Reactor or TSR.

Although there are simplified procedures of temperature scanning (TS) possible in certain specific cases, their applicability and the chance of introducing unknown errors by their use make them less than attractive in most cases. In general, the procedures described in Chapter 5 should be scrupulously followed. However, an example of the justifiable use of a simplified TS procedure is given in Chapter 11. The temptation to use such simplified temperature scanning procedures should be avoided, because the method can lead to unappreciated errors and disenchantment. Considering the productivity of a TSR operated in the prescribed manner, it would be hard to justify cutting corners in any but the most robust cases.

Once we take the "great leap forward" in data collection using the TSR, we are faced with massive amounts of data and inadequate methods for handling this information. The first issue is the statistical cleanup of the data to remove noise, so the data from the TSR can be used effectively. In collecting data using the TSR no repeat readings are taken at any point, and consequently the error in the data is more like noise in electrical signal processing than it is like the familiar form of error commonly studied using repeat measurements at a single point. The way of dealing with this new sort of noise involves *filtering*, a mathematical procedure designed to smooth curves without distorting their underlying shape. It is the shape of the curves recorded by a TSR that contains the message that conveys information about reaction rates and thence about the reaction mechanism. Some of the text below outlines the state of our current understanding about the use of filters in the case of TSR data.

The second issue involves mechanistic rate expressions that are often devilishly hard to distinguish from one another, even on the basis of masses of very good data. Currently, statistical methods are available to guide conventional experimentation to regions of reaction conditions where rival models are most readily discriminated. Similar procedures can be applied to TSR data but are less useful in this application. What is needed is an extension of the available model-discrimination methods to TSR data. Some consideration has been given to this but much remains to be done.

The final problem is the most troublesome. In fitting candidate rate expressions to TSR data, one has to start the iterations with a set of initial parameter values. It would be simple if all parameters could initially be set to a value of one to start the iterations. This will not work with available parameter optimization routines. The problem is serious and the search for good starting values is empirical and potentially lengthy. One often zeroes in on a set of optimum parameters only to find that further searching, from a new set of starting values, can improve the fit. Even the correct criterion of fit is poorly understood. An unweighted sum of squares of residuals is commonly used but its universal applicability is much in doubt. The difficulties are connected with the shape of the surface being investigated in the parameter space, rather than with the existence of local minima. These, by and large, are the result of errors associated with the data. In principle, a well-filtered set of TSR data should yield a smooth surface, yet a sure way of zeroing in on a unique set of parameters from arbitrary starting values continues to elude us.

The problem of finding a global minimum in parameter fitting is partly due to a disparity in the relative size of the components of the kinetic rate constant and partly to the inherent problems of fitting sums of exponential terms to data. Perhaps there is no solution and we will always be troubled by this issue. However, now that the data is available and the need for better fitting procedures is before us, there is an incentive to improve parameter optimization routines to reduce the subjectivity present in these procedures. Then again, subjectivity in this type of data-fitting may well be inevitable.

Encouragingly, some fixes are already in hand. Routines with automatic scaling and step size adjustment are the simplest remedial procedures and are often applied. The final answer may lie beyond this, in the development of algorithms better suited to the optimization of parameters using the data, rate equation forms and parameter spaces that delineate reaction kinetics. All these can now be well documented by the use of the TSR and it remains to be seen if data processing can rise to meet the challenge of improving the fitting of this data to mechanistic rate equations.

It seems that good judgment and intuition in the processing of kinetic data will never be replaced by strictly objective methods that are capable of full automation.

1. Reactor Types and their Characteristics

The mark of a true professional is a thorough understanding and mastery of the tools available to do the job.

A Broad Classification of Reactor Types

Investigation of reaction rates can be carried out using a number of distinguishable types of reactor. Each type has its advantages in specific circumstances, and the proper choice of reactor-type will have a major impact on the cost, speed, ease and success of the investigation. It is therefore important to know how to choose the right reactor type for a given investigation.

Two broad classes of reactors can be identified: batch reactors and flow reactors. The two types also involve two types of operation: transient and steady state. In some reactor configurations, aspects of more than one type of reactor are incorporated into one unit, a practice that should be carefully examined case by case for its applicability to the work to be done and its ability to deliver reliable and valid data. Such hybrid reactors are in general tricky to operate and have a limited range of accessible operating conditions. Although they can be built to simulate plant operating conditions, they are usually a poor source of kinetic data, particularly if it is to be used to understand the dependence of the reaction rate over a broad range of reaction conditions or to quantify the reaction mechanism.

The Batch Reactor

Configuration

The batch reactor (BR) is the most intuitively obvious reactor configuration. Much exploratory chemistry is carried out in such reactors, especially at the early stages of the development of a new synthesis. The familiar laboratory beaker into which reagents are introduced to carry out a reaction is perhaps the most commonly used BR. Such simplicity is attractive, but when we come to apply a BR to kinetic studies, additional sophistication must be introduced. Most importantly, we must make sure that temperature and composition are uniform throughout the volume of the reactor. In the commonly employed method of isothermal BR operation, vigorous stirring and temperature control are employed to make sure that the temperature and composition are uniform at a predetermined level throughout the volume of the reactor.

The need for isothermality requires a means of adding or removing heat from the vessel at a rate sufficient to maintain a constant temperature in the face of the exothermic or endothermic heat effects due to reaction. The means of achieving this vary, from the use of internal heat exchangers, to external constant-temperature baths, to the recirculation of reactants through an external thermostatic device. The preferred method of temperature control depends on the amount of space available inside the reactor vessel for the heat exchanger, the amount of heat to be transferred, and the heat transfer possible through the exchanger walls to the thermostatic medium.

If heat transfer through the reactor walls is adequate, the simplest temperature control method is to use an external bath or jacket that contains a cooling or heating medium whose temperature is controlled by means of heaters or refrigeration. For this method to work well, the thermal effects of the reaction have to be relatively small, heat transfer through reactor walls must be easy, and both the reacting mixture and the heat exchange medium must be stirred vigorously so that contact with the heat transfer surface at the reactor wall is maximized. If only heat is to be added to the reaction, the reactor vessel can be placed in an electric heating blanket and the temperature maintained by controlling the power supplied to the heater. However, there is a down side to this simple design, in that temperature overshoots in exothermic reactions carried out at elevated temperatures are hard to control, due to the inadequacy of the cooling surface area. This problem can be overcome by introducing an internal cooling coil into the vessel, but at that point the reactor design and the temperature control procedure become cumbersome in terms of both control and mechanical design.

A second problem with this type of arrangement is that reactions requiring thick-walled vessels, such as pressure vessels, are subject to serious lags in response to the control inputs. The reactor therefore tends to cycle between a high temperature and a low one, reducing the reliability of the kinetic data. There are measures that can be taken to reduce this cycling. A proper control policy is helpful, but its development may require extensive preliminary runs and, useful though it may be in an industrial reactor, it is usually not justified in a laboratory investigation. Placement of the temperature sensor can present a better, if only half-measure, alternative. For example, placement of the sensor at (or even in) the reactor wall may give better response and control in a well stirred vessel than placement in the middle of the reacting mixture.

Yet another problem in using an isothermal BR in kinetic studies arises when the reaction is to be carried out at elevated temperatures. If the reactants are placed in the vessel and then raised to reaction temperature, an unavoidable fraction of the charge will be converted during the temperature-raising process. The reaction therefore cannot be observed in its initial stages at the reaction temperature being investigated. Since the behaviour of reactions near (in principle, at) zero conversion is very important to the understanding of reaction mechanisms, this deficiency presents a major disadvantage of the BR.

The optimal design of a laboratory isothermal BR would therefore consist of a well insulated closed pressure vessel with a stirrer, an internal electric heater, an internal cooling device, and a thermal sensor properly placed inside the vessel. The stirrer, cooling/heating surfaces and the sensor would require careful positioning to achieve uniformity of temperature and composition, but it can be done in a sufficiently large vessel. Unfortunately most research reactors are small and the above requirements mean that the BR is not often used in kinetic investigations. Its merit, when it can be used without encountering the above concerns, is that it is relatively simple to construct and operate and requires a limited amount of reactants.

In bimolecular reactions, a modified batch reactor can be used to introduce a second reactant in such a way that its concentration remains constant with time. An example is a reaction between a liquid and a gas. In such cases the liquid reactant, and any catalyst that may be required, are placed in the batch reactor while the gas phase component is sparged through the reacting mixture. This method allows a sufficient amount of gas to be introduced to carry the reaction to any required level of conversion. Delaying the introduction of the second component can be used to delay the start of the reaction until the desired reaction temperature is reached.

This type of operation involves sophisticated control and analysis methods. Reactor operation and the interpretation of the results are complicated by the need to measure and account for the almost-constant concentration of the gas phase reactant as the liquid phase reactants are converted.

Modes of Operation

The principal requirement for the operation of a BR in kinetic studies is the availability of a rapid means of analyzing the reacting mixture without withdrawing a significant amount of the charge in the sampling process. These issues are important in BR operation because the BR is inherently a *transient* reactor. Once the moment is gone, there is no way to obtain a second sample at the same conditions, short of repeating the whole experiment. The preferred methods of analysis therefore involve *in situ* sensors such as electrodes or FT-IR cells. Failing this, mass spectrometric (MS) analysis, in which a minimal sample is aspirated out of the reacting mixture, can be applied. All methods of sampling are complicated if solids (such as fine catalyst particles), or dispersed gas bubbles, are present. These interfere with the sampling procedure by distorting the sample or plugging the sampling port.

Methods that require the withdrawal of large samples, or a long time for the analysis, complicate the use of the BR for kinetic studies and reduce the number of analyses obtained in each run. There is no way to control the speed of the reaction at a given temperature, and hence the number of points along a conversion vs. time trajectory is limited by the number of discrete conversion readings that can be taken during the course of the reaction. It is also limited by the number of samples that can be withdrawn without distorting the composition of the reacting mixture. All in all, the optimum analytical method for a BR used in kinetic studies is one that offers a continuous readout of the composition of the reacting mixture without withdrawing any material from the reactor. This type of analysis is in fact the ideal to which all kinetic studies aspire but, whereas in the operation of other reactor types there are ways to minimize or even eliminate this need, in the BR it presents a pressing necessity.

The other requisite is the minimization of conversion during the temperature ramping before the system arrives at the reaction temperature we intend to investigate. In cases of monomolecular reactions the solution is to make the temperature ramp as steep as possible. This presents the danger that, during the heat-up, the temperatures at the heat transfer surfaces will be much higher than the bulk temperature in the reactor. Such high temperatures will speed up reactions in the volume next to the heat transfer surfaces during the heat-up period. This could even lead to higher heat-up-time conversions than those that would have taken place at a slower heating rate and, most importantly, to distortions in the selectivity observed during the run, since various reactions in the overall process will be speeded up to a different extent.

In bimolecular reactions the two reactants can sometimes be heated up to reaction temperature separately and then mixed rapidly. This avoids the heat-up time but can be dangerous if the reaction is highly exothermic.

After all these requirements are met, the isothermal BR yields data consisting of composition versus *clock time,* i.e. the time since the reaction began at the desired temperature, composition and pressure. Successive runs can yield the same data at a series of temperatures, pressures and initial compositions. The course of the reaction during the clock time that elapsed between the start of the reaction and the time when it reached the desired temperature and we began to take readings remains unknown.

The Plug Flow Reactor

The plug flow reactor (PFR) can be envisaged as a conveyor carrying microscopic batch reactors from inlet to outlet. The erstwhile reaction time that we measured as clock time during the course of a reaction in a BR is now replaced by *space time,* which is a measure of how long it takes an increment of feed (the microscopic BR) to travel the length of the conveyor/reactor at a constant speed.

The great advantage of this arrangement is that once the flow rate and temperature are stabilized and invariant operation is achieved, the composition at the outlet of the reactor does not change with clock time. The PFR is therefore a steady state reactor. This allows us to sample repeatedly at a given condition and verify our results. In terms of BR operation, it is as if we had frozen the progress of the reaction at some level of conversion and can now allow ourselves unlimited quantities of sample for analysis as the product issues out of the PFR. Arguably, this is the principal reason for the dominance of the PFR over the BR in kinetic studies. This ideal situation is achieved only under certain conditions. Care must therefore be taken that the PFR is built and operated so as to approach the required conditions as closely as possible.

Configuration

The laboratory PFR normally consists of a constant-diameter tube made of suitable material and surrounded by a temperature controlling device. In order to achieve the desired level of conversion the volume of the reactor tube must be of appropriate size compared to the feed rate. This relationship is fundamentally expressed as space time

$$\tau = V_{reactor} / f_0 \tag{1.1}$$

where τ is the space time in units of time;
$V_{reactor}$ is the reactor volume in units of volume;
f_0 is the volumetric feed rate in units of volume occupied by the feed at inlet conditions, introduced into the reactor per unit of time.

This brings up the first problem: the definition of the feed rate varies widely and is often not reported in sufficient detail. Various measures of feed rate result in different measures of the space time, the essential measure of the time that an increment of feed spends in the reactive volume. In fundamental studies the dimensions of the feed rate should always be convertible to units of time spent by a mole of the reactant in the reactive volume. This time must be such that it can be related to the clock time the material would have spent in a BR operated under the same reaction conditions. Only in this way can we obtain reaction rates that are comparable between investigations done in those two reactor types. Even for more utilitarian purposes, where the use of fundamental units is of lesser importance, one should establish a well defined measure of the feed rate if comparisons between studies done in different reactors and laboratories are to be possible.

Despite this, and for a variety of reasons, the definition of space time is not always as clear as one would like and researchers often settle for some fairly arbitrary, often murky, quantity to represent the space time. This leads to differences in reported rates that make reliance on literature values of rates, not to mention the utility of rate parameters, quite unsatisfactory. For example, in a PFR packed with inert material, the volume of the reactor in space time calculations should be the void volume between the packing particles, not the volume of the empty reactor. It is not clear that this is the definition used in all cases of reported data.

In heterogeneous catalytic reactions, the problem is even more subtle: what exactly is the reaction volume? If the catalyst has a porous structure accessible to the reactants, is it the bulk volume of the porous catalyst particles? Or perhaps the volume of the pores themselves? What if the catalyst is non-porous and reaction takes place only on the external surfaces of the non-porous particles? Should we then use the void volume or the catalyst volume as the reaction volume? Or perhaps all catalytic reaction rates should be expressed per unit of time per unit of available surface area.

$$\tau = S_{catalyst} / f_0 \tag{1.2}$$

where τ is the space time in units of active area of catalyst * reactant volume^{-1} * time leading to overall units of length^{-1} * time;
$S_{catalyst}$ is the total active area of the catalyst charge present in the reactor;
f_0 is the volumetric feed rate in units of feed volume *at inlet* conditions.

In practice, in catalytic reactions, space time is often defined using the weight of the catalyst rather than its volume or surface area. This is not only easily measured but, if the proper information is available, can be readily translated to volumetric or surface terms. A consequence of this convention is that the units of space time are now:

$$\tau = w_{catalyst} / f_0 \tag{1.3}$$

where τ is the space time in units of catalyst weight * reactant volume^{-1} * time;
$w_{catalyst}$ is the weight of the catalyst present in the reactor;
f_0 is the volumetric feed rate in units of volume occupied by the feed at inlet conditions.

Often the feed rate is also reported in units of weight. This is fine as long as the molecular weight of the pure feed is known, since weight can then be translated to any suitable units. Confusion can arise if the molecular weight of the feed is not known.

However, even then a reproducible result can be reported for comparison purposes, if the expected molecular weight of the feed is the same in the studies being compared.

The use of these and various other units for the space time requires the use of correspondingly altered units for the catalytic rate constant. The rate constant can now include the mass of the catalyst, and also be dependent on catalyst bulk-density, active site surface-density, porosity and accessible surface area. Catalyst properties in terms of these quantities must be known in detail before valid comparisons between competing catalysts can be made.

Most of these questions are not important beyond the problem of comparing results from different laboratories, since many of the space time definitions vary from each other by just a calculable constant. However, in some cases other issues may be involved, such as the assumed volume of vaporized liquid feed at inlet or at STP conditions, the definition of catalyst density, and so on, making comparison between results from various laboratories difficult and raising the potential for distorting the calculated activation energies and frequency factors of a reaction. There is no generally accepted convention for the definition of space time and the best we can expect is to know exactly how it has been defined in each case. *There is an urgent need for this information and it must be demanded of authors by referees and editors if we are to remove some of the fuzz from kinetic data in the literature.*

Modes of Operation

The PFR is conventionally operated at isothermal, almost-isobaric, conditions. Moreover, it must be operated at flow velocities high enough for turbulent flow to be present along the length of the reactor. This is required to eliminate radial gradients and make the reactants move along the reactor as radially uniform "plugs". This requirement makes the PFR approximate the conveyor of microscopic BRs, as described in previous discussion. Velocities that are adequate for turbulent flow in a tubular PFR can be estimated from the expression:

$$N_{Re} = Lv\rho/\mu > 2100 \tag{1.4}$$

for empty tube reactors or from:

$$N_{Re} = D_p v\, \rho/\mu > 40 \tag{1.5}$$

for packed bed, such as catalytic, reactors. The notation in the above is:

- N_{Re} is the Reynolds number
- L is the length of the reactor
- D_p is the diameter of the particles in the packing
- v is the linear velocity of reactant flow based on the empty cross section of the reactor (at reaction conditions)
- ρ is the reactant density (at reaction conditions)
- μ is the reactant absolute viscosity (at reaction conditions)

These limits give the minimum velocities necessary for plug flow operation. Operation at lower Reynolds numbers can be expected to result in a radial profile of velocities in the reactor, with the result that the time of transit of increments of reactant varies along the radius of the reactor cross section. This in turn means that a sample of effluent from the reactor contains increments with different transit times whose *average* transit time is the *nominal* space time for that run.

It can readily be seen that such an average transit time does not guarantee that the conversion of the effluent, which is an average conversion due to the various transit times resulting from the distribution of radial velocities, is the same as that to be expected if all radial increments had the same transit time, in each case equal to the space time. Thus, there is a lower limit to the flow rates applicable to a given reactor and packing, before it can be considered to be operating as a PFR. Slower flow rates should not be expected to produce conversions corresponding to those expected from the micro BR on our imaginary conveyor.

There is also an upper limit to the flow rates, and hence a lower limit to space times accessible in a given PFR. This limit is more intuitive and arises because of the pressure drop that takes place when fluids flow through a tube, be it packed with solids or empty. The pressure drop can be calculated, but for purposes of kinetic experiments, the pertinent question is: what pressure drop can be tolerated? In the study of reaction rates of incompressible reactants and products, the answer is that very large pressure drops can be tolerated since such reactions show little sensitivity to pressure, and the incompressibility of the fluids involved guarantees that flow is uniform throughout the reactive volume regardless of pressure gradients.

This is not so if one of the reactants or products is compressible. In that case both flow velocity and all concentrations change with pressure as well as temperature. As flow proceeds through the reactor and pressure drops, the volume of the compressible constituents increases, flow velocity increases, and the concentration of the reactants decreases. The same effect takes place if there are temperature gradients along the reactor and, unavoidably, if there is a difference between the number of molecules of products and the number of reactant molecules that formed them. The simplest relationship that accounts for these effects is in the form of a linear expression for the volumetric flow rate of the reacting mixture in gas phase:

$$f_i = f_0\left(1+\varepsilon X_i\right)\frac{P_0 T}{P T_0} \tag{1.6}$$

where f_i is the local volumetric flow rate of the gas stream
P is the local pressure in the reactor,
T is the local temperature in the reactor,
X is the mole fraction of the reactant converted at that point in the reactor,
f_0 is the volumetric flow rate of feed at 0 conversion and STP,
P_0 is the pressure at STP,
T_0 is the temperature at STP,
ε is the volume expansion/contraction coefficient defined as:

ε = [moles of products at 100% conversion - moles of feed at 0 % conversion]/ [moles of feed at 0 % conversion]

The consequence of the assumed linear dependence of expansion/contraction on the various conditions is that concentration terms in the rate expression interpreting data from a PFR must be corrected as follows.

$$C_i = C_0 \frac{1-X_i}{1+\varepsilon X_i}\frac{PT_0}{P_0 T} \tag{1.7}$$

The correction for expansion therefore depends on the stoichiometry and the reaction conditions. The above correction factor is simplified if we operate at constant P and T conditions along the reactor axis. Even more simplification comes by assuming at the same time that the standard condition (subscripted 0 in the above) is that at the entrance to the reactor. In that case the only factors affecting the concentration are the degree of conversion, X, and the stoichiometric expansion coefficient ε.

$$C_i = C_0 \frac{1-X_i}{1+\varepsilon X_i} \tag{1.8}$$

From the original relationship, it is clear that a pressure drop across the catalyst bed will affect reactant concentration by a factor of (P/P_0). This can result in significant distortions to the kinetic interpretation at various flow rates, since various pressure drops are used to vary space time in an experimental program. At what point this becomes a problem is a matter for the experimenter to determine. Clearly, at some flow rate, and hence at some pressure drop, the distortion becomes unacceptable. This defines the upper limit of accessible flow rates.

In Chapter 7 we will consider this problem in more detail. In particular we will examine methods of correcting for volume expansion in temperature scanning reactors, where temperature and conversion vary significantly during each experiment. We will see that the conventional use of the linearly dependent epsilon factor to account for volume expansion is by-and-large unsatisfactory.

The Continuously Stirred Tank Reactor

Configuration

The continuously stirred tank reactor (CSTR) is identical in design requirements to the BR except for the introduction of a constant flow of feed and a constant withdrawal of reactor contents, so that the reactant volume in the reactor remains constant. The CSTR is therefore inherently a *steady state* reactor. This is accomplished by keeping the active volume, temperature, feed rate and pressure of the CSTR constant. If anything, the mixing in a CSTR must be even better than in the BR since an ideal CSTR is one in which the input stream is *instantaneously* mixed with the reactor contents so that the stream being withdrawn is identical in composition with that of the reactor contents. Besides the need for efficient and rapid mixing, the CSTR requires the same considerations regarding temperature control and the design of reactor internals as does the BR. In gas phase reactions the CSTR also exhibits volume-expansion effects identical to those encountered in the PFR. These require correction to the concentration terms as described above for the PFR and as will be considered in more detail in Chapter 7.

What is unique in the operation of the CSTR is that conversion is taking place throughout the volume of the reactor at constant composition. This means that the CSTR *directly* yields the rate of reaction at the constant composition of the reactive volume. There is no need to take slopes or process the readings in any way but by inserting them into the rate expression. All we do is simply measure the difference in conversion between the inlet and the outlet and divide that by the space time, defined in any one of the ways it was for the PFR. This gives rate measurement in a CSTR a great advantage over rate measurements in a PFR or a BR, where in one case we must take differentials between analyses from several separate runs at different space velocities, and in the other, make separate analyses of several samples taken within a short interval of clock time. Since the CSTR is a steady state reactor, we can readily take a number of samples of output composition at steady state to make sure that the change in composition between the reactor inlet and outlet is well established.

One would think that the ability of the CSTR to deliver rates of reaction directly would make it the reactor of choice in research, but it is not so. It should certainly be so for homogeneous liquid phase reactions; there is little or no reason to shy away from the CSTR in such investigations. In gas phase reactions, and in particular in heterogeneous catalysis, the situation is less promising. The reason lies in the relative complexity of the required reactor internals, including the difficulty of designing a means of catalyst retention while making sure that there is efficient contacting of the gas phase with the solid catalyst. In particular, these requirements make it difficult to design a conveniently small laboratory CSTR.

Although small CSTRs are available on the market, their ability to operate with gas phase reactants over a broad range of conditions is not well established. As a consequence, catalytic CSTRs tend to be one liter or larger in volume, require heavy containment vessels if they are to operate at elevated pressures and temperatures, and involve elaborate seals to allow access for a stirring shaft to enter the pressure vessel. All in all, catalytic CSTRs tend to be heavy, complicated and expensive. Moreover, the presence of adequate mixing is often under debate.

There are numerous ways that one can attempt to do the mixing in a catalytic CSTR. The most obvious and the most certain way is to construct a reactor that consists of a catalyst bed through which all the reacting gases are rapidly recirculated by means of a positive displacement pump. Into this loop we would introduce a relatively small stream of fresh feed and, somewhere further along the loop, after the catalyst bed and well away from the inlet port, we would withdraw the output stream. The whole apparatus would have to be maintained at reaction temperature and pressure, and in many applications the pumping mechanism would therefore operate in difficult conditions. This requirement makes this straightforward design difficult, and consequently it is rarely used in practice.

The other seemingly simple design would involve the dispersion of the catalyst in the reacting fluid. This approach may be successful in industrial-scale reactors, but at the scale of laboratory reactors, separation of solid catalyst from the fluid in the exit stream is not easy. This problem has rarely been solved in a way that allows both the desired small scale and a minimal holdup in the separation device.

The result is that most of the common CSTR designs involve high speed turbines that operate inside the reactor volume and are designed to force the reactant fluid through a thin bed of catalyst. Two commonly used design approaches have emerged: the Berty reactor, one of whose embodiments consists of a high speed turbine drawing the fluid through a thin horizontal bed of catalyst situated in a draft tube above the turbine rotor; and the Robertson reactor in which the gases are drawn or forced into the center of an annular cylindrical basket containing the catalyst. The fluid is then forced out radially through the walls of the annular basket. In both designs the recirculation is internal to the volume of the reactor, eliminating problems of temperature control and pressure containment. The remaining problem of driving the turbine is solved by the use of magnetic drives that transfer the torque available from rotating magnets driven by an external motor to magnets on the turbine shaft contained in a non-magnetic pressure-containing housing, operating at reactor pressure.

There is a third design, the Carberry reactor, which is well known and attractive in its conceptual simplicity. Unfortunately, it has proven less attractive in the light of its operating problems. This reactor consists of a radial or cruciform wire-mesh basket containing the catalyst, immersed in a vessel that defines the reaction volume. The catalyst basket is mounted on an axially located shaft that is rotated rapidly so that the basket arms are forced through the reacting fluid. Mechanical and fluid-dynamic problems have prevented adoption of this design as a standard for CSTRs. It remains to be seen if a successful variant of this design can be engineered.

Steady State Operation

Conventional flow reactors operate at steady state. This requirement involves the stabilization of the composition of the reacting mixture and of the temperature of the mass of the reactor vessel and, in the case of CSTRs and BRs, of the reactor internals. The achievement of this condition usually requires long periods of stabilization before a steady state is assured. It is not uncommon for a CSTR to take a day to reach stability at a new reaction condition. The situation may be somewhat better in the case of changes in feed rate and/or reactant composition without a change in reactor temperature, although in principle all transients in CSTRs decay exponentially and take forever to complete.

In all steady state flow reactors, the operating policy is to wait for steady state to be established before a valid reading is taken. The presence of the steady state is detected by observing an output condition as a function of clock time after the imposition of a change in operating conditions. To make sure that the criterion of compositional steady state is fulfilled, the approach is to take periodic samples of the effluent and analyze their composition. A frequently used and simpler shortcut is to wait for the temperature of the reactor to stabilize, at which point it is assumed that compositional steady state has also been established. This procedure is convenient and labor saving, but should be checked in each investigation by the more certain procedure of taking samples for compositional analysis before accepting temperature as a proxy indicator of steady state.

Once more, a rapid and preferably continuous method of analysis makes the identification of steady state, as well as repeat analyses at steady state, more practical. The taking of a single sample at an assumed steady state, not uncommon in view of the length of analysis time in many systems, can lead to exaggerated scatter in conversion data and consequent uncertainties in the calculation of reaction rates. These in turn lead to uncertainties in the form and parameters of the rate expression that is fitted to the data, making it less useful in theoretical and practical applications.

Other Reactor Types

Besides the three classical reactor types, there are numerous reactor configurations and modes of operation that enlarge the range of methods of data collection in kinetic studies. Each of these is supported by a more or less adequate understanding of the mixing processes involved and by specialized methods of interpreting the data collected. In most cases such specialized reactors are limited in the range of operating conditions that can be studied, or operate at conditions well removed from those of commercial interest.

Among reactors that are designed to examine the elementary processes occurring on catalyst surfaces, the Temporal Analysis of Products (TAP) reactor (Gleaves, J.T. et al., 1988) presents an interesting development. This and a number of other reactor types depend on transient response to reveal details about the overall kinetics. Among the other transient reactor types that examine changes in conditions during reaction are those employing Temperature Programmed Desorption (TPD) (Amenomiya, Y. and Cvetanovic, R.J., 1963) and Temperature Programmed Reaction (TPR), types in relatively common use. Data obtained from these reactors is at best semi-quantitative and therefore useful only for comparative purposes, not for kinetic studies. In cases that do aspire to quantitative interpretation, the mathematical procedures used for data interpretation are either very complex or highly simplistic, with the result that the conclusions obtained are often shaky and not very informative. The equipment required to operate such reactors also tends to be complex and expensive. A review of some of these methods is available (Bennet, C.O., 1976).

A few reactor types other than the above-mentioned ones attempt to deal with specific problems caused by the physical characteristics of the reactants and catalysts involved. Below we briefly examine two such configurations.

Fluid Bed Reactors

Two types of fluidized bed reactors can be distinguished: confined bed and flowing-bed reactors. The first type is used when the catalyst is so fine that plug flow through a fixed bed of this catalyst would present an unacceptably high pressure drop. Instead, the catalyst is fluidized by an upward flow of feed. It is maintained in the fluidized state, but not carried over out of the reactor, by careful adjustment of the feed rate. Temperature control in such reactors is very good and the pressure drop negligible. The problem is that beds of this type offer only a narrow range of feed flow rates between full fluidization and unacceptable carry-over of catalyst with the product stream. This limits the range of space times that can be studied and makes the fluid bed reactor inappropriate for broad ranging kinetic studies. In principle this deficiency can be overcome by using a series of fluidized reactors, each capable of allowing fluidization at a different range of feed flow rates and hence at different space times. This solution is less than satisfactory; normally,

extrapolations of conversion vs. time behaviour toward conditions of zero conversion but only after steady state is established. Such procedures can yield initial rates for the steady state phase of the mechanism but do not solve the problem of measuring the rates of the establishment of the mechanistic steady state. The study of these details of a mechanism is a problem that remains in the realm of specialized reactor techniques.

High Throughput Screening Reactors

The advent of combinatorial methods of catalyst formulation has led to the development of a wide variety of methods for screening large numbers of samples for catalytic activity. Devices designed to do this display a broad range of physical configurations, mostly clustered in the region of small scale, even micro, reactors. High throughput screening (HTS) reactors involve exclusively BR or PFR type configurations and are rarely designed to measure reaction rates. In the cases where rates are measured, a great deal of thought has to be given to the errors that may be associated with ultra-small scale of operation and other features of these devices before accepting the measured rates of reaction as valid for purposes of mechanism interpretation or even for comparison to values obtained using other reactor types. On the other hand, the ingenuity of some of the new configurations and the impressive automation of the hardware is attractive, and many of the design and automation features that have appeared in HTS reactors can usefully be introduced in reactors designed for kinetic studies.

General Thoughts on Reactor Configurations

Chemical reaction takes place whenever the reactants are placed in conditions where the reaction *can* take place. As a result reactions will take place and can be observed in detail in a variety of vessels, under a great variety of operating conditions, using many contacting patterns. In various specific cases, reactors that include highly idiosyncratic features have been built and used for testing of reaction properties. These many reactor types serve a purpose and provide reproducible results which then serve to guide research in that specific topic. However, the requirements that must be met in order to obtain *valid kinetic data* limit the types of reactor vessels, contacting patterns, and operating conditions that can be used to no more than a handful. The simplest way to see this is to consider the fundamental case of the BR. In order for the rate of reaction to be measured correctly the BR must be:

- of constant volume during the reaction;
- constant composition throughout the volume of the reactants;
- at a constant, uniform, temperature throughout this volume.

In this volume we must be able to measure the composition as frequently as possible and maintain homogeneity. A fairly simple mathematical treatment of the composition data, outlined above, allows us to extend this simple case and deal with a BR whose volume changes with conversion.

It is these same requirements that have to be met in all reactors designed for kinetic studies, with the added problem of finding an appropriate definition of space time so that the results obtained in flow reactors designed for kinetic studies can be compared to those from a BR. These requirements exclude most reactor configurations from use in kinetic studies and leave us with the fundamental trio: the BR, the PFR and the CSTR. Other configurations can yield reliably reproducible data but fall short in one way or another when used for kinetic studies.

The preceding discussion is intended to clarify the issues and to define the requirements of a reactor intended for kinetic studies. Gathering kinetic data in an inappropriate reactor configuration and attempting to relate this information to industrial design or to mechanistic studies is not recommended. A great deal of time and money has gone into generating confusion in this way.

2. Collecting Data under Isothermal Operation

There is a profound difference between "collecting data" and "recording measurements". The first makes available information in coherent sets which can then be used to construct scientific insights; the latter merely creates a catalogue of facts.

Collecting Raw Data

In experimental work on reaction kinetics three parameters are recorded:

- temperature (T),
- conversions or product yield (X_i) and
- time (t or τ).

Properly collected in adequate amounts, these three quantities, the T, X, τ *triplets*, provide all the information needed for identifying an appropriate kinetic rate expression.

Measuring the Reaction Temperature

The recording of temperature requires a calibrated sensor but otherwise is relatively straightforward. There are only two aspects that require care. The first is the location of the sensor. This has to be such that the measured temperature is representative of the temperature at which the reaction is taking place. A misplaced sensor could read a temperature distorted by, for example, the proximity of heat transfer surfaces or by the effects of mixing. The second is that the measured temperature must remain constant throughout the duration of observations and throughout the volume of the reactor. These latter aspects can be distorted by improper temperature control or by inadequate attention to heat losses and/or inputs in various parts of a reactor. Multiple temperature sensors may have to be located throughout the reactive volume to make sure that these possibilities have been eliminated.

Measuring the Reactant and Product Concentrations

The measurement of concentration presents a vast, active and challenging field of instrumental techniques of chemical analysis. This is the field whose development has the largest impact on kinetic studies. In principle there are two types of methods of analysis: batch and continuous. As a familiar example of a batch analysis we can take standard gas chromatography, where discrete samples are analyzed and require a fairly long time for an analysis to be completed. On the other extreme are methods involving electrodes and other probes that output an analogue signal that varies continuously with composition.

Between these extremes is a rapidly developing field of rapid batch analysis, where the time of each discrete analysis has been reduced to the point that the analytic signal is continuous, for all practical purposes. Such a method is found in mass spectroscopy combined with deconvolution of the mass spectra. This method combines frequent sampling of a multi-component stream with mathematical procedures that allow the rapid extraction of compositional information from otherwise uninterpretable overall mass spectra of product mixtures. Compared with the established gas chromatography-mass spectroscopy (GC-MS) methods of analysis for similar mixtures, the deconvolution procedure is two or more orders of magnitude faster.

All in all, development of rapid analysis is the limiting factor in the development of advanced kinetic experimental work and, by implication, in the availability of the masses of data required for the development of advanced kinetic interpretations. The development of robust remote IR sensors and the use of FT-IR methods plus the deconvolution of IR spectra of mixtures is the next major advance to be implemented in this field. The development of combinatorial methods of catalyst formulation and the rise of HTS methods of catalyst evaluation has provided a useful push to the development of new methods of rapid instrumental analysis.

The Definition of Contact and Space Time

The rate of a reaction is time dependent. It is therefore important that we know how to measure the period of time during which the observed change in concentration has taken place. Time may look like the easiest parameter to record, but this is not so. In fact the definition of time in kinetic studies deserves to be re-examined in much more detail.

The concept of reaction time is a trivial matter as long as the reaction is taking place in a BR at constant volume. The time pertinent to that system is the clock time, which begins at zero when the reaction is initiated. In the simplest case, if we mix two reactants instantaneously and start the clock at the same time, conversion in the BR will progress with clock time from the moment of mixing, and the time dimension on which the rate of the reaction depends is clear.

Consider, however, the situation where the reaction involves a change in volume. If the BR is of a constant-pressure design, then as the reaction proceeds concentration changes, not only due to conversion but also due to the change in the volume of the reactants. In a constant pressure BR this aspect is best treated by relating the volume of the reactants to the degree of conversion, as was described in Chapter 1, and including this in the kinetic equations. Clock time, however, would go on uniformly and the rate of reaction would continue to be a function of this time.

When we consider the same two cases as they appear in flow reactors the concepts of reaction time become less clear. In the case of a constant-volume homogeneous phase reaction taking place in a PFR, an increment of feed has a transit time, or space time τ, during which it traverses the length of the constant temperature plug flow reactor. Both these times are equivalent in this case. Assuming that the reactants were preheated in a relatively small volume and are cooled in a similarly small volume, the main course of the reaction takes place in the reactive volume of the reactor proper. The time spent in this volume can be calculated as the volume of the reactor divided by the volumetric feed rate of the reactants.

$$\tau = V_{reactor} / f_0 \tag{2.1}$$

where $V_{reactor}$ is the active volume of the reactor;
f_0 is the volumetric feed rate to the reactor.

The tau (τ), called space time, is a well defined quantity under these circumstances. In an ideal PFR, operating with one phase and no volume changes, it is the time that each increment of feed spends at reaction conditions in the active volume. It is obvious that in this case, for each increment of feed, the space time corresponds directly to the clock time of a microscopic BR carried through the reactor by an imaginary conveyor belt. Conceptually, and in reality in this case, the PFR is simply a method of transporting incremental BRs through the reactive volume in a time whose duration is controlled by the feed rate.

However, when volume expansion takes place, the flow rate is not constant throughout the reactor length, and as a result the defined space time is not equal to the real residence time, or transit time, of an increment of reactant. It is as if the conveyor belt changed speed along its length. Fortunately, as long as we can relate the changes in volume to the level of conversion, we can introduce this relationship into the rate expressions and thereby account for the effects of volume expansion on the residence time, albeit indirectly. This lets us use the same simply-defined space time τ as a measure of the time that we need to use to quantify the kinetics of the reaction. Thus, as long as we account for volume expansion by adjusting the concentrations in our kinetic rate expressions, we can use space time, defined at the entrance of the reactor, as the time pertinent to the time-differential in the rate of reaction.

This is not the end of the story. Space time is usually defined at room temperature and pressure, whereas the entrance condition to the reactor may be at a different temperature and a higher or lower pressure. Moreover, these conditions will change in subsequent runs as data is collected at several temperatures and pressures. The effect of using space time defined at room temperature and pressure for all such runs, without taking into account actual entry and reaction conditions, has an effect on the rate parameters of the fitted rate equations. It may even lead to differences of opinion as to the correct mechanistic rate expression that applies to a given reaction, if other workers define the space time differently. The same problems also lead to misconceptions as to the thermodynamic and kinetic values appropriate to the reaction, resulting in fundamental misunderstanding of the reaction itself. More obviously, this procedure can interfere with comparability between studies.

The use of standard temperature and pressure (STP) conditions for the calculation of the space time means that all kinetic parameters are evaluated as if the residence time of the reactants was that appropriate to STP conditions, regardless of the actual experimental temperature and pressure. The actual residence time is corrected, indirectly, using the volume expansion formulas cited above to account for changes in the concentrations of the reactants in the PFR. The rates thus measured in a PFR operating at a given temperature and pressure correspond to those that would have been measured in a BR whose volume was that of the feed at STP and whose operating conditions were the same as in the PFR.

Problems with the Definition of Space Time

Additional problems arise when one is considering a reactor packed with solids through which reactant is flowing. In the case of a homogeneous reaction in the presence of a non-porous packing, the reaction occurs in the interstices between the particles of the packing, and all we need to know is the total volume of these interstices. However, if the interstices are porous and have significant pore volume, the real contact time becomes hard to define. Such packings should not be used for kinetic studies of homogeneous reactions. Unfortunately, this is usually exactly the situation we face in the study of heterogeneous catalysis.

Begin by considering a catalyst whose active surface is all on the exterior of non-porous particles. Reaction therefore takes place only on surfaces exposed in the interstices of the packing. What volume should we use for calculating the pertinent time for the measuring the rate of the reaction? Neither the volume of the interstices nor that of the solid packing correctly represents the active volume in the reactor. It is the area of the active surface exposed to the reactants that is the best representation of this volume. We are forced to consider a very different measure of space time:

$$\tau = S_{catalyst} / f_0 \qquad (2.2)$$

where $S_{catalyst}$ is the active surface area in the volume of the reactor
f_0 is the volumetric rate of feed supply to the reactor

This time τ is in units of (time area volume^{-1}) or (time length^{-1}) and will generate a frequency factor in the rate expression that will contain these unusual units. There is no easy way around this kind of units unless one is willing to say that the external surface area of the catalyst corresponds directly to the bulk volume of the solid catalyst. Such a relationship will depend on the particle size for small particles of non-porous solids.

As it happens, a simpler situation, where surface area is proportional to the weight or the volume of the solid, is well approximated in the case of highly porous catalysts. In such catalytic reactions the space time can be calculated from:

$$\tau = V_{catalyst} / f_0 \qquad (2.3)$$

where $V_{catalyst}$ is the volume of porous catalyst in the reactor
f_0 is the volumetric rate of feed supply to the reactor

Other definitions of space time may be convenient but are sure to introduce factors that will distort rate parameters and reduce comparability between studies. Since the literature is full of idiosyncratic definitions of space time, and since the best fit rate expressions and rate parameters can be influenced by these definitions, one should check this issue thoroughly before comparing conclusions in different literature sources.

Identical strictures apply to the CSTR, despite the fact that few increments of feed spend the calculated space time in the reactor. In fact the distribution of residence times is exponential and, in principle, some increments will take forever to exit the reactor while others exit as soon as they enter. Nevertheless, the distribution of residence times is such that the defined space time is correct for treating kinetic data obtained in a CSTR as long as appropriate corrections are made to the concentration terms.

These difficulties with the definition of reaction time lead to frequency factors that are difficult to predict or reconcile with physical realities on the basis of fundamental concepts that will be discussed in Chapter 9. The only solution is therefore to make sure that the definition of space time is clearly reported and that the units used can be changed by the reader to other units using clearly understood conversion factors.

Often the term *space velocity* is used in the kinetic literature. This is the inverse of space time and, as long as it is properly defined it too can be used in kinetic rate expressions. Unfortunately, most of the time this quantity is even less suitably defined and a kinetically useful space time can only sometimes be calculated from reported space velocities. Much data present in the literature is therefore lost to kinetic studies.

The Dimensions of Rates

The rate of conversion in a chemical reaction involves a change in the number of moles of the chosen component as a function of time. The time dimension has already been discussed. We now look at the question of the quantity of material converted.

Rather than dealing with the extensive quantity of the total number of moles of material under consideration, it is convenient to consider an intensive quantity, say the number of moles per unit of volume. For this reason rates are generally examined as changes in concentration per unit of time. The concentration can be expressed in a number of ways, of which the most common are:

- moles per unit volume of the reacting mixture;
- partial pressure of the component;
- weight per unit volume of the reacting mixture.

Other units can be imagined and a great variety of units is to be found in the literature. This variety notwithstanding, in order to reduce experimental data to useful kinetic parameters, the units of concentration have to be in terms of molar concentration per unit volume. Chemical conversion affects individual molecules of the reactant, not units of its weight. Other units that may be reported in the literature must be reduced to molar units if one intends to use mechanistic rate expressions and/or arrive at thermodynamically meaningful constants in fitting rate data.

This implies that the analytical data collected must report changes in molar concentration or some quantity readily convertible to this measure. Measures of peak height or signal intensity and other such measures of signal strength coming from an analytical instrument cannot be used directly for kinetic interpretation unless they are linearly related to the molar concentration of the component. On the other hand, any such signal can be used after re-scaling, no matter the complexity of the relationship, as long as there is an established unique connection between the signal and molar quantities. Between the acquired raw data and the requisite concentrations lie various transforming relationships; sometimes several such steps are required. Often the molar quantity reported in the raw data is the mol fraction of the component present in an output stream. Methods of relating those quantities to concentrations and their use in kinetic rate studies will be discussed in Chapter 7.

As an example of a transformation of raw data to kinetically usable concentrations, consider measuring the concentration of product in hydrolyzing acetic anhydride. This can be done by measuring the conductivity of an aqueous solution of hydrolyzing acetic anhydrate. The raw conductivity and the concentration of the free protons formed by hydrolysis are related by:

$$C_{H+} = [(T_t - T_0)/(T_\infty - T_0)](2K_aC_{A0})^{1/2} \tag{2.4}$$

where T_t is the conductivity of the solution observed at time t
T_0 is the conductivity of the solution observed at time t=0
T_∞ is the conductivity of the solution observed at time t =∞
K_a is the equilibrium constant for the ionization of acetic acid
C_{A0} is the initial concentration of acetic anhydride in (mole liter^{-1})
C_{H+} is the concentration of acidic protons during the decomposition

The kinetics of this reaction can be readily analyzed after the appropriate conversion of the conductivity signal to concentration terms, but cannot be determined directly from the conductivity signal.

Similarly the concentrations of individual components in the exit stream as analyzed by mass spectrometry may or may not be correctly represented by the intensity of the parent peak of the said component. Two steps are required to transform the raw data to kinetically useful quantities. The first step stems from the fact that each mass peak may contain contributions from a number of components in the effluent. To resolve this difficulty, a mathematical procedure called *deconvolution* is applied to the observed total mass spectrum of the exit stream from the reactor. Deconvolution allows us to separate the total intensities at individual mass numbers into their several components, each of which is due to a contribution arising from the spectrum of an individual constituent present in the effluent sample. The deconvoluted spectrum then reports the mol fractions of the individual components in the effluent. With this in hand, the next step is to transform the output mol fractions to concentrations so they can be used in rate expressions to correlate reaction rates.

Data Processing and Evaluation Methods

The raw data obtained by experiment contains *scatter*, a distortion of numerical readings that can be ascribed to random events in the procedure, equipment function, and human error. It is taken as an article of faith that, if the equipment is correctly designed and operated, random errors are reduced and the remaining errors will be normally distributed around the true value. On the other hand, incorrect experimental procedures or bad equipment design may also produce normally distributed error, but centered on a false value. Other sources of error can skew the distribution of readings so that the arithmetic average is displaced from the true value, invalidating the commonly applied averaging procedure. These systematic errors can be hard to spot, identify, or correct and they can be seductive in the resulting appearance of normalcy.

The sources of experimental error are rarely examined in detail and the statistical treatment of observational error is generally simplistic and based on convenient assumptions. It is therefore of primary importance to design, construct and operate equipment in a way that is free of systematic error and whose random error is minimized to the extent possible, all within the limits of available and affordable technology. Having said that, error is a fact of experimental life and must be dealt with.

The simplest procedure, and the one almost universally applied, is to perform repeat runs at a set of conditions to establish the presence and extent of (presumably) random error. The arithmetic average of this set of repeat readings is assumed to be a better approximation of the true value. Unfortunately, each reading adds a substantial unproductive burden to the experimental program so that repeat runs are few in any investigation. The preferred procedure is to perform a few runs at one convenient experimental condition, and to calculate some measure of error dispersion at that point. Since there are rarely enough repeat measurements at any other point to do a serious evaluation of the error, a simplistic measure such as the average error is often based on the error estimate at this one point:

$$s_{av} = \Sigma \, |(c_i - c_{av})| \, / \, n \tag{2.5}$$

where n is the number of repeat readings;
s_{av} is the average deviation;
c_{av} is the arithmetic average of the n readings
c_i is the i'th reading in the set of repeat readings.

In cases where the absolute value of the readings c_i varies significantly from point to point, a similar formula can be used to calculate the percent error and applied to all points regardless of magnitude.

A more general method of evaluating error involves the collection of a sum of squares of residuals between a fitted function and all the data points involved in the fitting, repeated measurements and single measurements alike. The corresponding estimate of the magnitude of the error is now calculated as either the sum of squares of the residuals:

$$s^2 = \Sigma \, (c_i - c_{calc})^2 \tag{2.6}$$

or as the variance:

$$s^2 = \Sigma \, (c_i - c_{calc})^2 \, / \, (n - p) \tag{2.7}$$

where p is the number of parameters being fitted.

It must be noted that this result, and indeed the whole investigation, is invalid if $p>n$. This will be obvious when we note that this is simply another way of saying that a minimum of n data points is required to obtain a unique set of parameters for an n-parameter function.

These various sources of error make it difficult to obtain satisfactory raw rates of reaction. There is no way of eliminating all error in the X, τ, T readings from conventional reactors. As a result, the rates of reaction obtained by taking discrete differentials of raw conversion data with space time (in the case of the PFR and the BR) are subject not only to the primary error in experimental readings of X and τ, but also to the amplification of this error by the taking of discrete $\Delta X/\Delta\tau$ slopes from such noisy data. More will be said about noise reduction in Chapter 7.

Converting from Concentrations to Mole Fractions

One normally analyzes the effluent stream in terms of mole fractions of components present at the reactor outlet. This allows us to perform mass and atomic balances on the effluent and thereby ensure the consistency of the data. On the other hand, mechanistic rate expressions are normally developed in terms of concentrations. What is required is therefore a method of transforming concentration-based rate expressions to a fractional conversion form. This is particularly important in treating data from a temperature scanning reactor (TSR) (see Chapter 5) but is important in dealing with the analytical data from most reactors before the data can be fitted to mechanistic rate expressions. The procedure for doing this is well defined. In liquid and solid phase reactions, where there is no change in volume with reaction, the problem can be ignored; in the gas phase one proceeds as follows.

Take the general stoichiometry for a bimolecular reaction:

$$aA + bB \leftrightarrows cC + dD$$

where the capital letters stand for molecular species and the lower case letters indicate the moles of each required by the stoichiometry. The first choice is that of the *base component*. The stoichiometry is then normalized on this base component.

$$A + b/aB \leftrightarrows c/aC + d/aD,$$

where the quotients are the stoichiometric coefficients for the case of the selected base component and will be called ν_i.

Since concentration at any local flow rate depends on the moles flowing per unit time divided by the volume passing by per unit time, the local concentration is:

$$C_i = F_i/f \qquad (2.8)$$

As conversion progresses this concentration changes. For the base component it is:

$$C_A = F_A/f = F_{A0}(1-X)/f \qquad (2.9)$$

and for all other components it is:

$$C_i = F_i/f = (F_{i0} + \nu_i F_{A0}X)/f, \qquad (2.10)$$

where ν_I is by definition positive for products and negative for reactants. Notice also that the fraction converted is defined as $X = (F_{A0} - F_A)/F_{A0}$. This is usually not available directly from the mol fraction of A at the outlet of the reactor. More will be said about this in Chapter 7.

This general formula does not provide us with a method of calculating the local flow rate f. To calculate this we first calculate the number of moles in the reacting mixture after converting one mole of the base component. This turns out to be

$$\sigma = \Sigma \nu_i \tag{2.11}$$

Using this *expansion factor* we calculate the ratio of the change in moles on total conversion of one mole of the base component divided by the total moles of feed to the reactor. The total moles at the output are:

$$N_T = N_{T0} + \sigma N_{A0} X \tag{2.12}$$

$$N_T / N_{T0} = 1 + \sigma (N_{A0} / N_{T0}) X \tag{2.13}$$

This allows us to define the *expansion coefficient* ε

$$\varepsilon = \sigma (N_{A0} / N_{T0}) \tag{2.14}$$

From the ideal gas law we write:

$$V = V_0 (P_0/P)(T/T_0)(N/N_0) \tag{2.15}$$

If we accept N to be N_T, the total moles in the reacting mix at time t, the above equation becomes:

$$V = V_0 (P_0/P)(T/T_0)(1 + \varepsilon X) \tag{2.16}$$

Instead of the volumes V and V_0 we can write the above equation for volumetric flow rates f and f_0. This lets us write the concentrations in terms of inlet conditions and the expansion coefficient:

$$C_A = F_A/f = F_{A0}(1-X)/(f_0(P_0/P)(T/T_0)(1 + \varepsilon X)) \tag{2.17}$$

$$= [(1-X)/(1 + \varepsilon X)](F_{A0}/ f_0)(P/P_0)(T_0/T))$$

$$= C_{A0}[(1-X)/(1 + \varepsilon X)](P/P_0)(T_0/T))$$

and

$$C_i = F_i/f = (F_{i0} + \nu_i F_{A0} X)/ f_0(P_0/P)(T/T_0)(1 + \varepsilon X) \tag{2.18}$$

$$= [(F_{i0}/ F_{A0} + \nu_i X)/(1+ \varepsilon X)](P/P_0)(T_0/T)$$

$$= C_{A0} \ [(y_{i0} + \nu_i X)/(1+ \varepsilon X)](P/P_0)(T_0/T)$$

where y_{i0} is the ratio of mols of component i to the base component, in the feed. The formulations above, or appropriate similar formulations in other specific cases, are substituted into the mechanistic rate expressions to transform these rate expressions

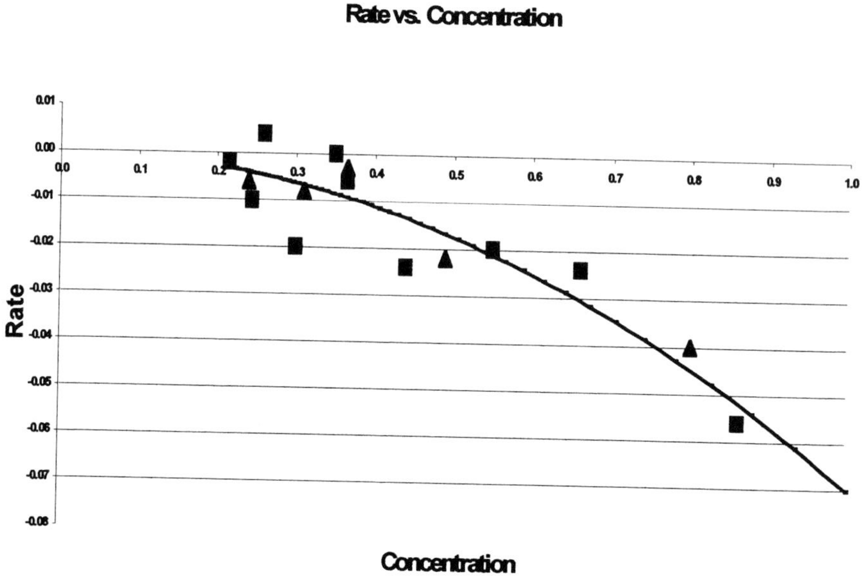

Figure 2.7
Comparison of scatter in calculated rates from experimental data obtained using the BR or the PFR. There is little to distinguish the scatter obtained from these two reactor types. Triangles present the BR data while squares are for the PFR.

Extracting Rates from CSTR Data

The operation of a CSTR is in many ways similar to that of the PFR. Feed is introduced at a constant rate and the product is analyzed after steady state is reached. Although the time to reach steady state is considerably longer than it is for the PFR, let us assume that we will take the same number of samples as we did in the PFR. For purposes of this comparison we will assume that the samples are taken at the same level of conversion as in the previous example and with the same errors. In that case the conversion vs. space time plot will look as shown in Figure 2.8.

Notice that although the scatter is similar to that in Figure 2.5, the space time scale is much longer. This means that either the volume of the reactor must be much greater or the feed rate must be much smaller than was the case for the PFR. Since the stirring is mechanical we do not have any pressure drop or Reynolds number limitations so that in principle we can cover the same range of conversions as before. But, whereas we were able to achieve this range of conversions within a tenfold range of feed rates in the PFR, now we find we need a range of thirty or more. This puts greater demands on flow controllers and, together with the slow rate of stabilization after the imposed flow rate changes between runs, places a time constraint on the use of the CSTR over this range of conversions.

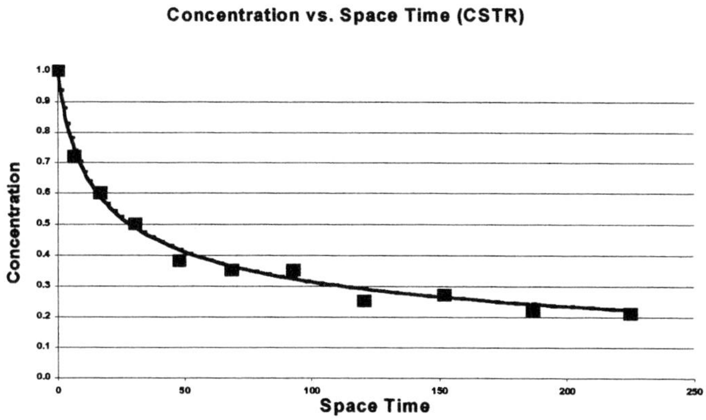

Figure 2.8
The true behaviour of a second-order reaction and a set of experimental points that might be observed as the system is investigated experimentally in a continuously stirred tank reactor

Figure 2.9 shows a compensating virtue of the CSTR. The rate of reaction in a CSTR is calculated from the expression:

$$r_{exp} = (C_{A0} - C_A) / \tau \tag{2.20}$$

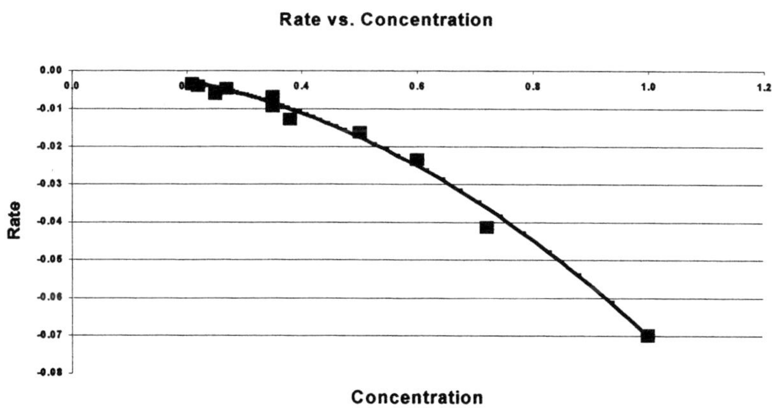

Figure 2.9
True second-order reaction rates and the experimental rates calculated from the CSTR data shown in Figure 2.8 by taking rates from equation 2.20.

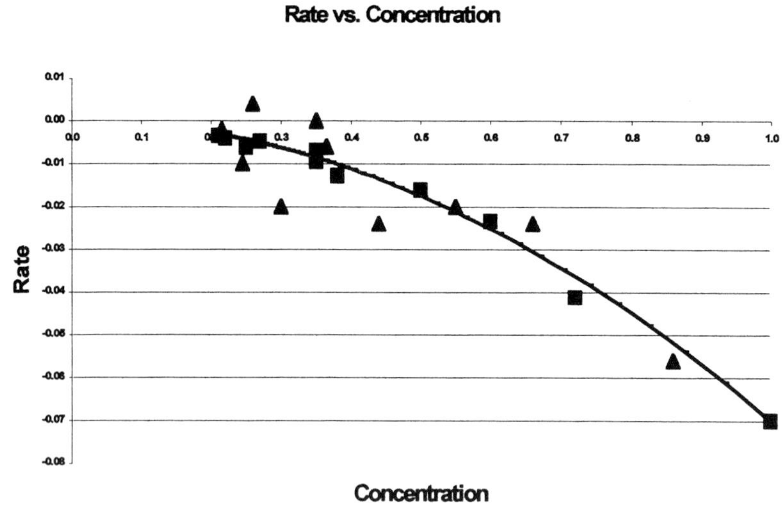

Figure 2.10
Comparison of scatter in calculated rates from experimental data obtained using the CSTR or the PFR. There is a significant reduction in scatter when a CSTR is used. The triangles are for the PFR and the squares for the CSTR.

This means that there are no approximations of slope or estimates of the corresponding concentration necessary. The rate is obtained directly from one error-prone measurement: the composition of the output. The other quantities in the rate equation are (in principle) well established by measurements and calibrations done before the experiment is started: volume of the reactor, volumetric flow rates reported by the flow meters, and the composition of the feed. This will reduce the scatter in the rate measurements, but even neglecting this advantage, the direct measurement of rates yields increased precision in the rate estimation as reflected in the plot of rate vs. concentration shown in Figure 2.9. Comparison between the scatter for the same range of conditions from a PFR, shown in Figure 2.5, and the results from a CSTR shown in Figure 2.9, is made clear in Figure 2.10.

We see at once that the advantages of using the CSTR for kinetic studies are substantial in terms of the quality of rate data to be expected, albeit at the expense of long stabilization time and more demanding volumetric flow measurement/control. Whereas the data from the CSTR might give us a glimmer of the underlying rate expression, conventional treatment of the data from the PFR holds out no such promise in our example. This is not to say that the PFR will not yield useful kinetic data. It will. However, the issues raised above point to serious difficulties in doing kinetics using conventional laboratory reactors. The ideal situation would be to acquire much more data than is usual, a goal that is hard to realize using any conventional laboratory reactor configuration, due to the need to establish steady state at isothermal conditions before a valid point can be measured.

Summary

A summary of the merits of the various reactor types begins by first noting that the BR has little merit as a device for rate measurements. Its hoped-for simplicity is often not possible to implement in practice and the limitations of data acquisition are serious. The flow reactors are clearly to be preferred and allow us to measure conversion at a greater number of reaction times and conversions. They also allow duplicate measurements to be made at any point. Unfortunately, the need for steady state to be established makes the acquisition of data in flow reactors tedious and time consuming. Moreover, the collected data consists of independent measurements each with its own error, often collected days apart. This procedure will tend to increase the magnitude of the error and to complicate subsequent efforts directed at its removal from the raw data.

The result of all this is that experimental programs using flow reactors tend to be lengthy and result in data that contains significant errors, unavoidably introduced by the discontinuity of operation. This is further complicated by the repetitious operation of fairly complicated equipment, which induces a strong desire to be done with it and to proceed with data interpretation. All this bodes ill for the production of plentiful and reliable data required for model discrimination in the search for the correct, mechanistic, rate expression. Little wonder that the rate expressions normally fitted to kinetic data make no pretense at offering a mechanistic insight into the chemistry of the reaction. What follows in Chapter 5 is a methodology offering a convenient solution to this and other difficulties presently bedeviling kinetic studies.

3. Using Kinetic Data in Reaction Studies

Uninterpreted data is a neglected gem. Uncut and unpolished by the expertise of a craftsman, it is nothing but a stone devoid of brilliance or significance.

The Rate Expression

The rate of a reaction is governed by reactant concentrations and by physical parameters such as pressure and temperature, but how it is governed by these conditions depends on the *mechanism* of the reaction. It is the mechanism that lies behind the kinetic rate expression. Understanding the mechanism and its rate expression allows us to *engineer* the reaction by influencing elementary steps in the overall conversion process.

We note from the beginning that the mechanism of a reaction does not usually change with reaction conditions; it is a fundamental property of the reaction. This is also true for catalytic reactions on a homologous series of catalysts at similar reaction conditions. In a given setting, the rates of the elementary steps of a mechanism depend solely on the concentrations of the reactants (and active sites in the case of a catalyst) and the reaction temperature. Of these, temperature affects the rate parameters only. The behaviour of these is well understood from thermodynamic considerations and well described by the Arrhenius equation containing:

- a pre-exponential, temperature-independent term A, multiplied by
- an exponential term containing an energy term E.

This dependence is so important that we will examine it more closely later.

The concentration dependence of the rate expression has a more varied effect on the behaviour of the kinetics since it dictates the algebraic form of the rate expression. Identification of a rate expression that fits the experimental rate data and agrees with a plausible reaction mechanism for the reaction requires an extensive experimental investigation of reaction rates, and much thought.

In order to gather enough reliable rate measurements, the investigation should employ the most appropriate reactor type, as discussed in Chapter 1. The choice of reactor type depends on the issues raised there and on the chemistry and physical properties of the system under investigation. The ultimate goal is always to produce enough *reliable* data to make it possible to identify the *correct* mechanistic rate expression. The identification of the correct rate expression is not a simple matter and is highly dependent on the validity of the data, the amount of data, auxiliary information, and the insight of the investigator.

Formulating Kinetic Rate Expressions

There are two types of reactions: elementary and composite. The composite reactions consist of any number of steps, each of which is an elementary reaction. The elementary reactions of a composite reaction interact in reaction-specific ways, yielding the mechanism for the overall reaction. These interactions can differ from mechanism to mechanism even though the elementary reactions themselves remain unchanged. It is therefore the elementary reactions that are the fundamental components of reaction mechanisms and need to be understood first.

Formulating Elementary Rate Expressions

There are only two types of elementary reactions: monomolecular and bimolecular. The easier of the two to understand is the bimolecular reaction.

Bimolecular Reactions

A bimolecular reaction takes place when two molecules that will react collide with sufficient energy to overcome the energy barrier that lies between their reactant states and the product. This thermodynamic energy barrier, plus some unknown additional energy, is called the *activation energy*. The Arrhenius equation tells us that its size has an important influence on the rate constant and therefore the rate of reaction.

The other factor in the rate of reaction is the frequency of collisions between the two species. It turns out that this frequency is proportional to the product of the concentrations of the two reactants. Thus the rate of an elementary bimolecular reaction between A and B varies as the product of the concentrations of reactants:

$$-r_{A,B} \propto C_A C_B \tag{3.1}$$

and therefore

$$-r_{A,B} = kC_A C_B \tag{3.2}$$

where $-r_{A,B}$ is the rate of disappearance of the reactant A or B. Rates of disappearance of reactants are written with a minus sign by convention.

C_A is the concentration of the reactant A.

C_B is the concentration of the reactant B.

k is the rate constant.

The rate constant itself has a structure and consists of a *pre-exponential* term and the activation energy term as per the Arrhenius equation:

$$k = A\exp(-E/RT) \tag{3.3}$$

where A is the frequency factor

E is the activation energy

T is the absolute temperature of the reaction

R is the ideal gas constant in units that make the exponent dimensionless.

The size of the *pre-exponential* (frequency factor) of k will depend on the units of A. In turn A depends on the units of the rate as measured: that is, on the concentration units and time units as defined in the space time. This is where the definitions of space time, discussed in Chapter 2, have their influence on the rate constant.

The units in the exponential must be such that the exponent is dimensionless. They depend therefore on the absolute temperature and the units of R. Comparison of kinetic parameters between studies is therefore critically dependent on the reporting of well defined units used in making rate, time and concentration measurements.

For the fitting itself we begin by forming the (X, r, T) triplets. Rate expressions contain only these three variables, although more than one conversion X (or its corresponding concentration) may be involved in a single rate expression, as is the case in an elementary bimolecular reaction.

Limits on bimolecular frequency factors

There is one more consideration: in fitting experimental data with a proposed rate expression we must not accept the fit unless the rate constants that are obtained are realistic. This is a requirement forced upon us by the physical world, which constrains the parameters of realistic rate constants within certain limits.

The pre-exponential (frequency factor) A is constrained by the physical realities pertaining to the behaviour of molecules. Because molecules have dimensions and mass, these limit the size of the frequency factor. In homogeneous bimolecular reactions A represents the number of collisions that a molecule, on average, undergoes per unit time. This number has been calculated to be approximately 10^{14} cc mol^{-1} sec^{-1} for collisions between two atoms at room temperature. More complex reactants will have lower frequency factors, often by many orders of magnitude, but higher frequency factors for elementary bimolecular reactions are impossible. This upper limit provides an unavoidable test for the validity of experimentally determined frequency factors assigned to homogeneous bimolecular reactions. In all other cases, when the frequency factor of an experimentally determined rate constant can be expressed in the same units, the same upper limit must be obeyed. Upper limits can also be derived, in appropriate units, for reactions between fluids and surfaces and for the reactions of adsorbed species.

Limits on bimolecular activation energy

The exponential term in the Arrhenius equation is the factor that determines the fraction of these possible collisions that is going to result in reaction. The fraction arises because only some of the colliding species contain sufficient energy for reaction to occur. The experimentally determined quantity in this term is the activation energy E which obviously cannot be infinite. In fact it cannot be higher than some reasonable approximation of the bond strength of the weakest bond in the colliding species, otherwise one or the other molecule would fall apart at that bond, a process called unimolecular decomposition and not the bimolecular reaction we seek. Since the highest bond strengths in organic molecules are typically in the order of 300 kJ mol^{-1} one can expect the activation energy in bimolecular reactions to fall below 300 kJ mol^{-1}. This upper limit in the activation energy provides a second test for the validity of experimentally determined rate constants assigned to bimolecular reactions.

Unimolecular Reactions

Elementary monomolecular reactions are called unimolecular reactions. This subtle difference in nomenclature distinguishes a type of reaction that involves the decomposition of individual molecules in a single elementary step. To do this the molecules must have a minimum of energy in order to break or rearrange internal bonds, and this minimum energy means that their rate constants contain a finite activation energy, E. How this energy is acquired is a story in itself, especially since it is usually observed that unimolecular reactions are first order in reactant concentration. This well defined order has caused considerable puzzlement as to how the energy of activation is acquired and how the unimolecular reactions take place at all. It was already known that energy is transferred between molecules by collision processes, but these are inherently bimolecular and therefore second order.

The mechanism of unimolecular reactions

The answer to the energization puzzle turns out to involve a simple mechanism. Events proceed as follows:

a. two molecules of the reactant A collide and in the process exchange energy. As a result one of the molecules becomes more energetic and the other, less. Occasionally such a collision results in an energized species, A*, with sufficient energy to undergo a unimolecular reaction. We write this elementary process as:

   Reaction 1 $A + A \Rightarrow A^* + A$

b. an energized molecule of this sort has a finite lifetime while the internal energy is assembled at the *critical co-ordinate* so that the reaction can occur. However, during that lifetime the energized species can undergo further collisions. Since such highly energetic molecules are rare, most of their collisions result in de-energization. This process is written as:

   Reaction –1 $A^* + A \Rightarrow A + A$

c. however, some fraction of the energized molecules escape collision long enough for the reaction to take place. This is the truly unimolecular step in the process. We write this as:

   Reaction 2 $A^* \Rightarrow B$

This sequence of events constitutes a reaction mechanism. The solution of all such mechanisms depends on the steady state assumption. This states that the net rate of change in the concentration of an unstable intermediate of a reaction is zero. To put this into mathematical notation we write:

$$dC_{A*} / dt = k_1 C_A^2 - k_{-1} C_A C_{A*} - k_2 C_{A*} = 0 \tag{3.4}$$

where the subscripted rate constants correspond to the reactions described above. Note that the net rate of change in the concentration of A* is an algebraic sum of the rates of its formation (positive) and rates of its disappearance (negative).

The steady state expression is next solved for the concentration of the unstable species in terms of rate parameters and the concentrations of stable species.

$$C_{A*} = (k_1 C_A^2) / (k_{-1} C_A + k_2) \tag{3.5}$$

Next we introduce the expression for this concentration into the rate expression for the conversion of reactant (or of the production of product):

$$-r_A = -dC_A / dt = dC_B/dt = k_2 C_{A*} = k_2\, k_1 C_{A2} / (k_{-1}\, C_A + k_2) \tag{3.6}$$

We see that at low reactant concentrations where $k_{-1}\ C_A << k_2$, the overall rate is in fact second order in C_A and depends on the rate of energizing collisions. When $k_1 C_A >> k_2$ the rate becomes first order in C_A and the experimentally observed rate constant is not an elementary rate constant but a composite of three elementary rate constants.

$$k_{exp} = k_2\, (k_1/k_{-1}) = k_2\, K_{en} \tag{3.7}$$

where K_{en} is the energization equilibrium constant.

Limits on unimolecular frequency factors

This mechanism has been confirmed by numerous experimental studies. Nevertheless, in the vast majority of cases unimolecular reactions are observed in their first order regime. The experimental rate constant in these cases is composed of an experimental pre-exponential ($A_{exp} = A_2 A_1 / A_{-1}$) and an experimental activation energy ($E_{exp} = E_2 + E_1 - E_{-1}$). Theoretical considerations lead to the conclusion that the pre-exponential of k_2 has an upper limit of about 10^{13} sec^{-1}, which corresponds to the vibrational frequency of an average bond that may break in the reaction. At the same time the frequency factors for k_1 and k_{-1} are expected to be very similar, making the expected $A_{exp} \approx A_2$.

Limits on unimolecular activation energies

Similarly, an upper limit of 300 kJ mol^{-1} can be expected for the experimental activation energy required to break the critical bond. While k_{-1} requires no activation energy, as it involves the loss of energy in a collision, the unimolecular decomposition, reaction 2 (with rate constant k_2) takes place without any activation energy since the species A^* already contains the required energy. The required energy is acquired by collisions in process 1 and will not be much greater than the energy corresponding to the bond strength of the bond that is about to be broken. Thus the experimental activation energy for this process will roughly correspond to the bond strength of the weakest bond in the molecule. These two limits offer a check on the validity of experimentally derived first order rate constants for elementary (unimolecular) reactions.

Reaction 5 $H\bullet + C_2H_5\bullet \rightarrow C_2H_6$

Applying the steady state approximation to all radical species, the unstable intermediates in this case, we calculate the steady state concentrations of these species in terms of rate parameters and feed concentration.

$$C_{CH3} = 2k_1/k_2 \quad (3.20)$$

$$C_{C2H5} = (k_1k_4/k_3k_5)^{1/2} C_{C2H6} \quad (3.21)$$

$$C_H = (k_1k_3/k_4k_5)^{1/2} \quad (3.22)$$

On the assumption that most of the conversion takes place in the propagation reactions we write the rate of conversion of ethane as:

$$-r_{C2H6} = k_3C_{C2H5} = k_4\, C_{H\bullet}C_{C2H6} = (k_1k_3k_4/k_5)^{1/2}C_{C2H6} \quad (3.23)$$

Surprisingly this complex sequence of events results in a first order rate expression for the overall conversion. However, the experimental rate constant of this rate expression consists of a square root of a quotient of products of elementary rate constants. This fact notwithstanding, the experimental rate constant is usually subject to limits in A and E, similar to those described above for elementary rate constants. For example, an analysis of the experimental frequency factor shows why it should be in the expected range of magnitudes. The experimental frequency factor is:

$$A_{exp} = (A_1A_3A_4/A_5)^{1/2} \quad (3.24)$$

Assuming that the individual elementary reaction frequency factors composing A_{exp} are limited in their magnitude to that theoretically possible, we see that the experimental frequency factor can be expected to be about the size of an average elementary frequency factor. For it to be unexpectedly large, the A_5 would have to be unusually small, while the other frequency factors would be normal or large. On examining the reactions that are involved there is no reason to expect this to be so. The same reasoning leads us to expect the overall activation energies of the composite rate constant to fall within the theoretically admissible limits for elementary reactions.

What is unexpected about rate expressions in chain mechanisms is that the order of the reaction, together with product selectivities, can change over a range of reaction conditions, *with no change in mechanism.* This happens when some reaction in the mechanism is superceded by a competing reaction. Such a transition can take place due to change in reaction temperature or pressure, or due to the addition of an inhibitor or catalyst. For example, it can be shown that if the termination reaction in the above mechanism becomes a $\beta\beta$ process, the order of the overall reaction changes to 3/2, while for $\mu\mu$ termination it becomes 1/2. If such changes take place over the range of temperatures under investigation, the reaction order displays intermediate values as the mechanism transitions from one order to the other, suggesting that an empirical power-law rate expression may be appropriate in this regime. This kind of behaviour accounts for much confusion in kinetics, and for many a failure to assign a mechanistic rate expression to data from experimental observations.

In cases where the kinetics is being observed in the transition region, the power exponent of an empirical rate expression may well be a function of temperature, further complicating the interpretation. In this situation treating the overall reaction as a network, with all significant competing reactions included, can be the appropriate approach, assuming that enough information about the individual elementary reactions is available from ancillary studies of the component elementary reactions. Alternatively, a more thoughtful approach would be to write the overall rate expression in the transient region as a sum of contributions from the rates from the two adjoining limiting regimes, and let their relative proportions be a function of temperature.

Whatever the procedure adopted in practice to deal with such contingencies, the fact is that there is always a unique set of reactions that lead to the observed behaviour of the overall kinetics. A full understanding of the reaction mechanism, and of the elementary reactions involved in it, provides the information necessary to engineer the reaction to a useful end, on a rational basis. Kinetic studies are an important tool in discovering this mechanism.

Among the various non-catalytic reaction mechanisms, chain mechanisms are particularly susceptible to engineering by means of interventions in the elementary processes constituting the mechanism. This allows the reaction engineer to adjust both the overall rate and selectivity for individual products. The enhancement or suppression of the rate of just one of the elementary steps in the mechanism can have a significant effect on these features of the overall reaction. Perhaps the simplest example of such engineering is the introduction of an inhibitor to suppress the chain transfer reaction by trapping out the initial active intermediates. Since initiating events are rare, a small amount of inhibitor can reduce the rate of reaction to a very low level.

A thorough understanding of the mechanism and kinetics of a chain process is essential to the systematic engineering of any chain reaction so that a specified purpose can be achieved. An empirical approximation of the rate of reaction simply does not supply the necessary information.

Catalytic Rate Expressions

The most effective method of altering the rate of a reaction is by using a catalyst. Catalysts alter the rates of individual reactions, or whole mechanisms, in ways that can be nothing short of startling. Reactions that do not proceed at measurable rates in the absence of a catalyst can be made rapid and selective by the introduction of minor amounts of a non-consumable material, the catalyst.

These remarkable materials owe their efficacy to their ability to form appropriate unstable intermediates which then channel the course of a reaction in ways that are improbable in thermal non-catalytic reactions. The action of catalysts therefore depends on a reaction between the reactant and an *active site* on the catalyst surface. Since catalysts can be solids whose activity resides on the surface, or homogeneously dispersed fluids, it is best to think of the active site as a reactive center on the catalytic molecule, even if the molecule turns out to be a crystal of solid material.

This concept lets us write the elementary mechanism of catalysis as follows. The reactant A is first adsorbed on the active site S to form an intermediate AS:

$$r_C = k_r\theta_A\theta_B = k_rK_AK_BC_AC_B/(1 + K_AC_A + K_BC_B + \sum_j K_jC_j)^2 \tag{3.32}$$

where k_r is now $k_2C_{S0}^2$. This is the *Hinshelwood Mechanism*. These three basic catalytic mechanisms formulated here constitute the building blocks of catalytic kinetics. More elaborate mechanisms contain combinations or elaborations of these three simple forms.

Even in its simple form, the Hinshelwood mechanism has some interesting properties. For example, it shows a maximum rate of reaction:

$$r_{Cmax} = (1/4)k_r \tag{3.33}$$

which occurs at $\sum_j K_jC_j = 0$ and $\theta_A = \theta_B = ½$. This in turn means that we require the reactant gas phase concentrations to be in the ratio:

$$C_A/C_B = K_A/K_B \tag{3.34}$$

and the surface to be fully covered, half by A and half by B. The appearance of a maximum rate of reaction at intermediate reactant concentrations provides an illustration of the surprising phenomena which make the establishment of the kinetics and mechanism of reaction of major practical importance.

Other limiting cases of these simple rate expressions can be formulated from any of the three basic rate expressions, but the general formula for the case should always be kept in mind. It can emerge in its more complete form in the middle of a range of conditions being investigated and make the experimental results deviate from a simplified version. Assigning such deviations to changes in mechanism is rarely, if ever, justified.

Properties of catalytic rate expressions

Catalytic mechanisms can result in very complex rate expressions that have an unfortunate mathematical form for purposes of data fitting. The various rate expressions that might apply to a given reaction are often hard to distinguish from one another on the basis of data-fitting. The reason is that many catalytic rate expressions can fit a given curve over a limited range of conditions. That, together with the fact that conventional studies yield sparse experimental results, leads to large uncertainties in distinguishing between rival rate equations. Even after a rate equation is chosen there are usually large uncertainties in the parameters established from the fitting of this sparse data with the model.

The only solution is to investigate broad ranges of reaction conditions and acquire massive amounts of data so that the behaviour of the kinetics is more broadly and more precisely documented. Failing that, one is often reduced to empiricisms and the use of a "standard run" for comparing the performance of successive catalyst formulations in a catalyst development program. As a result, despite good progress in applied catalysis, the understanding of catalytic reaction mechanisms is poor. This is highly unsatisfactory, considering the importance of catalysis to the chemical industry.

Enzyme catalysis

Besides the heavy chemical industry, where catalysis is a dominant feature of most conversion processes, enzyme catalysis is a critical component of bio-chemical processes. All that was said about mechanisms of catalytic reactions applies to enzyme catalysis. As can be expected, there are additional factors in enzyme catalysis that complicate matters. Many enzymatic reactions depend on factors such as pH, ionic strength, co-catalysts and so on that do not normally play a role in conventional heterogeneous catalysis. Despite this, the understanding of mechanisms in enzyme catalysis has outpaced that in heterogeneous catalysis and can now serve as a guide to the search for heterogeneous reaction mechanisms.

At the same time, heterogeneous catalysts have so many advantages in industrial applications that homogeneous catalysts such as enzymes are often "heterogenized" by attaching the active sites to a solid inert surface. The resultant catalyst usually differs somewhat from its homogeneous version but its mechanism of reaction should be very similar to that observed in the homogeneous form. Perhaps these are the systems that will serve as an entry to the systematic understanding of the mechanisms of heterogeneous catalytic reactions.

As we saw in the case of chain mechanisms, so in catalysis there are a number of families of catalytic mechanisms. Each family has its specific characteristics but all share some generalities. Among these is the formation of an unstable intermediate between the catalyst active site and the reactant. Understanding the nature and role of these intermediates is indispensable to the rational design of catalysts.

Solid-Gas Interactions

Non-catalytic reactions involving two phases are common in the mineral industry. Reactions such as the roasting of ores or the oxidation of solids are carried out on a massive scale but the rates of these processes are often controlled by physical, not chemical, effects. Reactant or product diffusion is the main rate controlling factor in many cases. As a result, mechanisms of reaction become "models" of reaction with consideration of factors such as "external diffusion film control" or the "shrinking core" yielding the various models. Matters are further complicated by considerations regarding particle shape and external fluid flow regimes.

Gas solid interactions are difficult to study systematically in conventional reactors but can readily be studied in a specialized type of temperature scanning reactor intended for this type of process, the stream swept reactor (SSR). In principle this is a batch reactor containing the solid through which the fluid phase flows sweeping out any desorbed material or reaction products to a detector at the outlet. Reactors of this type are also potentially applicable in adsorption studies and will be discussed in Chapter 5 under the heading TS-SSR.

Uses of the Mechanistic Rate Expression

Mechanistic rate expressions are by their very nature more broadly applicable than those formulated empirically. In principle a mechanistic rate expression should be valid over the entire range of approachable reaction conditions. Important as this may be, this feature is of little use in controlling an industrial process running at steady state, where departures from set point are normally small - hence some of the reluctance on the part of industrial researchers to pursue mechanistic studies. This neglect is not without cost, since mechanistic rate expressions can be of great utility at other than steady state operating conditions. The industrial utility of mechanistic rate expressions becomes important in two instances: during the process design phase, and in reactor control during startup and shutdown and other less foreseeable upsets. In the two cases of startup and shutdown, the breadth of reaction conditions that can be reliably quantified by the rate expression is crucial.

In the process of initial reactor design it may be possible to discover improved operating conditions by simulation during the design phase, if a broadly applicable model of the reaction is available. Traditionally, improvements in process operations come from actual operating experience, making full scale plants play the role of pilot installations, at considerable cost to the owner, as the operation of the reactor is incrementally improved on the basis of experience. Process improvements by these means may be achieved as long as the existing reactor and associated up-stream and down-stream processes can accommodate the changes that are found to be desirable. In cases where they cannot, expensive retrofits and de-bottlenecking are necessary, or the benefits of possible improvements do not appear until the next green-field version of that plant. Considering the lost revenue and opportunity costs involved in these procedures, it seems that a better understanding of the mechanism and kinetics of the reaction is very likely to be rewarding in terms of cost effectiveness.

During startup or shutdown, or potentially hazardous upsets in reactor operation, a full understanding of the reaction at conditions far removed from the design operating conditions is crucial. Of these two, the startup/shutdown transient is relatively routine and in general is handled well by empirically devised procedures. Although these may not be optimum in terms of productivity, they are at least reliable. Accidental transients are harder to anticipate. Their handling is often limited to an emergency shutdown with its associated costs and disruption. While an optimum method of handling of extreme transients using a full understanding of reaction kinetics may require standby resources such as emergency cooling, it would still be less disruptive than an emergency shutdown if reliable information about the behaviour of the system over a broad range of upset conditions were available and implemented in the control procedures.

Summary

Despite a prolonged lack of progress in the field, mechanistic understanding is recognized, sometimes implicitly, as the key to the engineering of reactions whose mechanisms involve combinations of interrelated reactions. The first step to such an understanding is to determine the reactions involved in the overall conversion process and their interdependences. This will identify the "handles" on the mechanism of the reaction. Then, if a method can be found to grasp one of the handles to alter the rate of an intermediate reaction in the mechanism, useful ends may be achieved. Searches for this elusive goal have been pursued empirically for decades, with some considerable success. Nevertheless, as in all of science, as the margin available for improvement narrows, fundamental understanding becomes essential to allow further progress.

New departures in catalyst formulation are also greatly facilitated by a thorough understanding of the mechanisms involved. If the mechanism of a reaction is well understood, focused efforts can be initiated in search of de-bottlenecking procedures to improve the pre-existing mechanism by altering the rates of intermediate steps on the sites of a new catalyst formulation. This is the rational design of catalysts that will make catalysis a science.

> *The failure of kinetics to fulfill its promise of serving as a guide to and validator of reaction mechanisms has long been a hobble on the progress of mechanistic studies. It has kept much of heterogeneous catalysis as an "art", preventing it from becoming a "hard science".*

4. Difficulties with Mechanistic Rate Expressions

The mechanistic rate expressions of some processes are inherently difficult to quantify. In such cases delineating the complexities of these processes with massive amounts of kinetic data may be the only way to arrive at a proper understanding of their mechanisms.

Problems of Parameter Scaling

All kinetic rate expressions contain quantities describing their dependence on temperature and reactant concentration. These quantities are badly mismatched in their influence on the rate of reaction. Particularly troublesome are the two quantities appearing in the rate constants, one of which describes the temperature dependence and exhibits exponential behaviour, while the other is a constant and differs from the unknown parameter in the exponential term by orders of magnitude. More specifically the two disparate quantities are connected by the relationship $k = A \exp(-E/RT)$. The frequency factor A is normally large, in the order of 10^{10} or so, while the activation energy E, in units of J/mol is only about 10^4. The upshot of this is that a unit of change in the frequency factor has much less effect on the size of the rate constant than does a unit change in the activation energy.

This mismatch causes poor estimates of rate parameters when fitting data unless one takes specific precautions. The situation is at its simplest when we are fitting an irreversible first or second order rate constant using the Arrhenius correlation. Using a proposed rate expression we fit data from several experiments and calculate the value of the rate constant at several temperatures. With this we make a plot of ln(k) vs. 1/T.

$$\ln(k) = \ln(A) - (E/R)(1/T) \tag{4.1}$$

This graphical procedure, which can easily be made analytical, provides a simple method of parameter scaling *and* data smoothing. Rate constants, obtained at various temperatures by fitting experimental data, are made available in the form of a numbers. These, when plotted on the Arrhenius coordinates, should *always* produce a straight line. There is good reason to suspect the validity of the rate expression if they do not. Statistical methods for dealing with such linear fits are readily available and yield the best fit frequency factor and activation energy. In simple rate expressions the scaling problem is therefore not overwhelming.

When the rate expression includes additive terms, each of which contains the above scaling problems, the situation becomes much more difficult. The fitting of data by a sum of exponential terms is a notoriously unsatisfactory procedure. Unfortunately this has to be done in many catalytic rate expressions as well as in many reaction networks. Although the problem of fitting this type of function has not been solved satisfactorily there are two established methods of dealing with it in a reasonably effective way:

- fit isothermal data sets separately and make an Arrhenius plot for each parameter in turn;
- fit all data at once with the full rate expression and use internal scaling of parameters.

Neither method is entirely satisfactory but, while the isothermal method is popular and robust, the method of internal scaling is more elegant and, in principle, more promising. Sometimes a third way of simplifying the problem is available by transforming the rate expression into a form that is less subject to scaling problems.

Transformation Methods

The known transforms of rate expressions are mostly based on attempts to linearize the rate function, say by inverting the function, so that the resultant form can be fitted using linear regression methods. Occasionally the rate expression can be put into a more convenient form by some other transformation which is applicable in that specific case, but there are no generally applicable transforms that make the job of parameter fitting easier or the iterations more efficient.

More inconvenient is the fact that all transformations distort the scatter of data about the fitted curve. In most cases the experimental error tends to be constant and is measured at points fairly evenly distributed over the range of experimental conditions studied. In fitting this data to a transformed rate expression, the commonly used statistical criteria of fit are invalidated by the fact that the transformed rate equation distorts the space defined by the reaction conditions and thereby changes the size of the error as it relates to the fitting of the transformed expression. This makes standard statistical methods used to obtain the confidence limits of parameter estimates, such as least-squares methods, less than robust and calls for data fitting routines that allow for an adjustment in the weight of the error over the range of the transformed parameter space. No general methods for doing this are conveniently available and few researchers use the methods that are known, because of their complexity. The final result is that the best fit parameters depend on the specific transform of the rate equation used in the fitting.

Scaling Methods

As was noted above, the rate constants $k = A \exp(-E/RT)$ are a source of particular difficulty in parameter estimation. Because A is normally so much larger than E, and because E appears in the exponent, when these parameters are fitted to experimental data, A and E exhibit a very strong and very non-linear correlation (see Figure 4.1). Taking the logarithm of k to linearize the relationship *á la* Arrhenius is not possible when there are sums of k's appearing in the rate expression. However, a simple scaling procedure that often improves the situation involves *temperature centering* about some temperature T_0 near the middle of the range of temperatures studied. To do this we rewrite the expression for k as:

$$k = A' \exp\left(-\frac{E}{R}\left(\frac{1}{T} - \frac{1}{T_0}\right)\right) \tag{4.2}$$

where
$$A' = A \exp\left(-\frac{E}{RT_0}\right)$$

This scaling procedure has the double effect of reducing the correlation between A and E and reducing the non-linearity of the correlation (see Figure 4.2). This in turn has two desirable consequences:

- the numerical least-squares procedures used to estimate the fit of the parameters will be more robust (i.e. less sensitive to poor starting estimates);
- commonly used statistical methods apply in a straightforward way to yield more-or-less valid confidence regions for the estimates.

A simple example illustrates the matter. More detailed discussions are given in Bates and Watts (1988) and in Watts (1994) where examples of parameter fitting are worked through in detail. Consider trying to fit the simple expression

$$k = A \exp(-E/RT) \qquad \text{(Model 1)} \tag{4.3}$$

to the following data:

T	480	500	530
k	0.838	2.382	27.355

The fitting will be done by finding A and E to minimize the root sum of squares

Model 1
$$S = \left[\sum_{i=1}^{3}\left(A \exp\left(-\frac{E}{RT_i}\right) - k_i\right)^2\right]^{1/2} \tag{4.4}$$

The global minimum is soon found at $S_{min} = 0.425$ for $E = 1.7507 \times 10^5$ and $A = 5.520 \times 10^{18}$. However, as Figures 4.1a and 4.1b show, the sum of squares surface for this simple system shows a long, curved, and very narrow valley near the minimum. This causes severe numerical difficulties in finding the global minimum and in estimating the confidence regions for the parameters. Indeed, using the conjugate gradient minimization algorithm (see Chapter 7) shows considerable sensitivity to starting estimates. For starting estimates close to the true minimum, the algorithm does indeed converge to the minimum, but for poorer starting estimates the algorithm converges to other points. For instance, starting at $E = 1.9 \times 10^5$ and $A = 20 \times 10^{18}$ leads to no change and the rather poor final estimates $E = 1.9 \times 10^5$ and $A = 20 \times 10^{18}$, with a sum of squares error of $S = 0.466$.

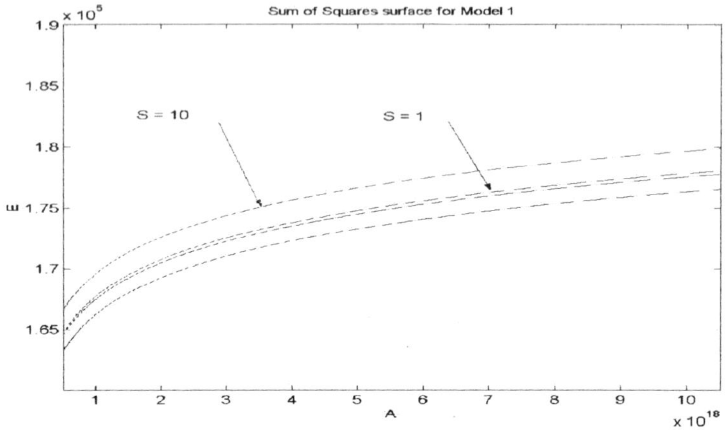

Figure 4.1a

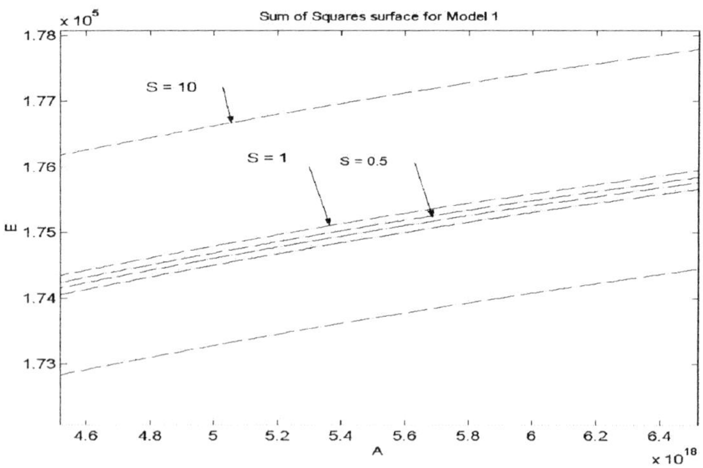

Figure 4.1b

Figures 4.1a and 4.1b.
Contours of the sum of squares surface for Model 1. For both figures the A and E ranges are centered at the global minimum. Figure 4.1b has an E-range 1/5 that of Figure 4.1a and shows that the problem of identifying the minimum persists as we approach the minimum.

A simple explanation of this is that the minimum is in fact, to a good approximation, any point on the curve along the bottom of the valley. Iteration while seeking the minimum leads us to a nearby point on the bottom of the valley, not to the global minimum.

If instead of Model 1 we try to fit a model centered at $T_0 = 500$, things are considerably improved. We now look for A′ and E to minimize

Model 2
$$S = \left[\sum_{i=1}^{3}\left(A'\exp\left(-\frac{E}{R}\left(\frac{1}{T_i}-\frac{1}{500}\right)\right)-k_i\right)^2\right]^{1/2} \tag{4.5}$$

Numerically, Model 2 is much less sensitive to starting estimates than Model 1. The global minimum occurs (as for Model 1) at $E = 1.7507\text{x}10^5$ and $A' = 2.50$ (corresponding to $A = 5.520\text{x}10^{18}$), with $S_{min} = 0.425$. For starting values of $E = 1.9\text{x}10^5$ and $A' = 0.248$ (corresponding to the starting value of $A = 20\text{x}10^{18}$ used with Model 1 above), the algorithm this time converges properly to the global minimum.

Figures 4.2a and 4.2b show the Model 2 sum of squares surface for the same ranges of E-values as Figures 4.1a and 4.1b. While there is still a large correlation between the estimates of A and E, it is much smaller than for Model 1. The fitting contours are much more elliptic, indicating a reasonably linear relationship between the estimates. Comparison of Figure 4.2 to 4.1 gives a clear picture of what scaling and rate expression transformations are meant to achieve: circular fitting contours with steep gradients surrounding the minimum.

The true frequency factor A can be calculated from the fitted pre-exponential factor A′ which, as it stands, does not have any physical meaning.

$$A = A'\exp(E/RT_0) \tag{4.6}$$

The value of E is obtained from the above fitting and T_0, the centering temperature chosen by the fitter.

One final comment about parameter scaling and transformation. In the case of fitting the simple expression

$$k = A\exp(-E/RT)$$

we have noted that fitting the log of k

$$\ln(k) = \ln(A) - E/RT$$

gives a linear expression in ln(A) and 1/T. The same simplification is not possible for rate expressions involving sums of such terms, but experience shows (see Watts, 1994) that fitting models using $k' = \ln(k)$ instead of k itself does tend to linearize the estimate correlations. Fitting ln(k) instead of k does not give as great an improvement as centering about T_0, but used in conjunction with centering gives some additional improvement. Fitting ln(k) also has the not-insignificant numerical advantage that the estimated values for k will automatically always be positive. This avoids having to impose artificial non-negativity constraints in the minimization algorithm.

is very difficult. The difference between the two mechanisms lies only in the mode of adsorption of oxygen: atomic in the first case, molecular in the second. Yet it is specifically this distinction that we want to establish in searching for the correct mechanism of the reaction.

Moreover, one could write similar mechanisms for the corresponding Rideal mechanisms where gas phase CO reacts with atomic or molecular adsorbed oxygen. One could also postulate that gas phase oxygen reacts with adsorbed CO, leaving an oxygen atom on the surface, and so on. The number of possible mechanisms for this simple chemistry is large enough to lead to considerable uncertainty as to the correct mechanism, unless massive quantities of good quality data are made available over a broad range of operating conditions so that a unique fit can be found. Even then, the "correct" set of rate parameters will be difficult to identify with certainty.

Transformations of Catalytic Rate Expressions

One way of getting an appreciation of which rate expression may be correct is to seek conditions where the general expression is reduced to simpler forms. For example, if we are able to achieve reaction conditions where the adsorption term for CO is much greater than that for O, the two above rate expressions reduce to:

$$r_{CO} = \frac{k_r K_{CO} K_{O_2}^{1/2} P_{CO} P_{O_2}^{1/2}}{(1 + K_{CO} P_{CO})^2} \tag{4.9}$$

and

$$r_{CO} = \frac{k_{r'} K_{CO} K_{O_2} P_{CO} P_{O_2}}{(1 + K_{CO} P_{CO})^2} \tag{4.10}$$

The rate expression is further simplified if the second term in the denominator is much larger than one, i.e. $K_{CO}P_{CO} >> 1$. In this simplified form one can see if the rate is dependent on the first power of oxygen concentration or on its square root.

The search for limiting forms of fully developed rate expressions is always useful if we know, or want to postulate, the full rate expression. But, if the rate expression being used is semi-empirical to start with, there is reason to believe that one is already dealing with a simplified, abbreviated, and perhaps distorted form of some unrecognized full rate expression. Changes in reaction conditions can now lead to seemingly new and unexpected forms of other semi-empirical rate expressions, suggesting a change in mechanism. This is almost sure to lead to confusion in any attempts to divine the mechanism of the reaction from kinetic observations.

At other times the operating conditions needed to reach some limiting behaviour are not achievable using the available equipment. The experimentalist in that case has to contend with extracting a broad understanding of the reaction kinetics from the data available. The kinetic behaviour under the available conditions may not be as simple as one would want. It is therefore best to seek procedures that are sure to lead to satisfactory fits in all cases and to support these with massive amounts of well distributed experimental data.

The corollary to the above simplifications is that simple rate expressions should be tested under conditions where they might depart from the simple behaviour initially observed. This search is equally important, though less often welcomed in practice. It is a truism that, when a rate expression departs from its simple form, it is a clear indication that the underlying mechanism is more complex than that initially envisioned.

This is generally true, since the mechanisms of reactions generally do not change with reaction conditions. Rather, the simplified rate expression of the overall process evolves smoothly from one form of rate expression to another due to changes in rate controlling steps of an ever-present network of reactions. All the simplified rate expressions are simplifications of some general rate expression, or reaction network, that encompasses all the sub cases. Once the limiting cases of the rate expression are understood, the full form, and often the underlying reaction mechanism, become apparent.

This, therefore, is the goal of kinetic studies: to investigate rates of reaction over a broad range of reaction conditions so that a unique and all-encompassing rate expression, corresponding to a plausible reaction mechanism, can be found.

The Integral Method of Data Interpretation

In some cases the proposed rate expression can be integrated to yield an analytical equation relating output conversion or concentration to the rate parameters. Data from isothermal studies is sometimes interpreted using this integrated form of the rate expression. The procedure avoids the need to take slopes in order to obtain rates from plug flow reactor (PFR) and batch reactor (BR) data, and may be an attractive alternative approach in cases where the integrated expression is easier to fit, or has a form that exhibits a lower correlation between parameters than the rate expression itself. In the best of cases it may allow a linear plot to be used for data fitting by regression. However, in the case of the temperature scanning (TS) operation (see Chapter 5), this procedure is of limited utility.

Limitations of the Method

In the case of data from a temperature scanning reactor (TSR), the integral method can be used when the temperature scanning reactor is operated in such a way that:

1. the space time in a TS-PFR is much shorter than the time it takes for a significant change in reactor temperature to take place;
2. in a temperature scanning batch reactor (TS-BR) the temperature change between successive analyses is small.

These two requirements present a relatively common situation and will present little impediment in practice. In the case of the TS-PFR it is also required that:

3. the temperature history of each increment as it passes through the reactor is known.

This is a more constraining requirement that is easily fulfilled only in thermo-neutral or isothermal reactions. If the operation is quasi-isothermal and condition 1 is fulfilled, the use of the integral form is greatly facilitated. However, it takes procedures described in Chapter 5 to interpret data in the case of reactions of high endo- or exothermicity where there is an axial temperature inhomogeneity along the TS-PFR.

The reason for the complications using the integral method in the case of the TS-PFR is that the integration of the rate expression requires full information about the temperature profile along the path of the integral. This temperature profile is unavailable from a typical TS-PFR run. Interpretation of data from this type of experiment is therefore limited to post experiment data processing. Only in the unusual case of an isothermal TS-PFR obeying condition 1 is it possible to interpret data using the integral equation in real time.

In cases where the integral method is applicable in a TS-PFR, one can sometimes calculate the rate constant with each analysis. This allows a real-time Arrhenius plot of the rate parameters to be generated as the ramping proceeds during an experiment. An example of such a procedure is presented in Chapter 11 where a study of the hydrolysis of acetic anhydride is presented. That reaction was carried out at high dilution and under conditions where quasi-isothermality of each increment was assured as it passed through the reactor. To make matters even simpler, the kinetics in this reaction are known to be first order, greatly simplifying data interpretation.

The required temperature information is, however, readily available in each TS-BR run, making the integral method applicable in that case. Whether the integrated form, using numerical methods and experimental temperature data, is to be preferred over taking slopes in the standard way that we will soon consider, remains to be seen.

The interpretation of TS-CSTR data in integral form does not come into consideration since that reactor makes rates available directly, without the necessity of taking slopes from surfaces. Integral methods can be used too, since reaction is taking place at isothermal conditions.

Caveat

Kinetic rate expressions are well known to exhibit hard-to-fit analytical forms. Moreover, most of them cannot be integrated to present a usable analytical form. We must therefore collect and fit data that reports instantaneous rates rather than cumulative concentrations. The use of kinetics to study reaction mechanisms is greatly hampered by these constraints. The only solution that can be envisioned is to acquire massive amounts of reliable, error-free, data. To achieve this we must clean up the raw experimental data by the skillful application of powerful methods of error correction. Only then is there the prospect that the data will reveal the underlying reaction mechanism. In the following chapters we present the necessary experimental methods for acquiring vast amounts of rate data and outline the early stages of the development of error correction techniques designed to deal with raw and noisy kinetic rate data.

5. The Theory of Temperature Scanning Operation

To change the status quo, you must challenge the current paradigm.

The Fundamentals

As noted in the discussion above, one of the main difficulties in identifying the mechanistic rate expression of a reaction lies in collecting enough good quality rate data to obtain a convincing and unique fit. Mechanistic rate expressions are often of a form that is difficult to fit with a high degree of certainty without collecting extensive rate data over a wide range of reaction conditions. Using conventional isothermal techniques, it takes a long time to collect a single data point. Much of this time is spent waiting for the reactor to reach isothermal steady state. Consequently, the acquisition of kinetic rate data is a notoriously laborious undertaking and the reluctance to collect enough data makes most of the fits available less than convincing.

The search for improved productivity in data collection has concentrated on reactor automation and the use of parallel reactors coupled to a single analytical system. This produces efficiencies and increases productivity from a given setup by a factor of five or so - too small to be revolutionary and often not very attractive in view of the cost of the automation and the added complexity of the equipment. Automated reactors also tend to be designed for a specific purpose, a feature that narrows the reactor's ability to serve in more than one investigation.

In the following we will examine an approach to kinetic experimentation in general that breaks through these restrictions. *The reactors to be described do not depend on a new configuration, rather they employ a new method of reactor operation.* The new method involves the *scanning* of reaction conditions and therefore collects data at non-steady state, as well as making full use of automation. This departure from standard modes of reactor operation results in a radically different method of data gathering and interpretation, rather than simply speeding up established experimental procedures. The resulting speed-up in data acquisition can be several orders of magnitude. As a result, the new methods promise to supply sufficient data, and data of better quality, so that fits of mechanistic rate expressions may be more believable. Most interestingly, the new methods promise to yield data that will allow sophisticated methods of error handling and removal, and to provide a new and significantly better method of accounting for volume expansion.

In the temperature scanning (TS) mode of operation the reactor is not required to be at an isothermal condition, and the reaction itself is not required to be in thermal steady state. The temperature of the reactor (and of the reactants) is made to vary during a run, in a controlled fashion at the inlet, and in a completely unrestrained fashion thereafter. In the particular case of the TS-PFR, not only will the temperature at the inlet and outlet change during a run, but there may well be a non-constant temperature profile along the length of the reactor. The thought that such a mode of operation can be used to gather valid kinetic data is startling.

Although the TS mode of operation does not require isothermal or steady-state conditions, it is assumed that the reaction is at all times in steady state with respect to certain steps in the reaction. For example, in catalytic TS-PFR operation, it is taken that the adsorption/desorption steady state is achieved much more rapidly than the time scale involved in the temperature scanning procedure. In the TS-CSTR we assume this, as well as the fact that complete mixing of reactor contents takes place on a time scale much shorter than the temperature ramping. Moreover, although there may be temperature differences and heat flows between various components of the reactor, of the catalyst, and of the reactants, these should not be flow-velocity-dependent, nor should there be any flow-velocity-dependent diffusion effects.

It is also assumed for now that there is no significant catalyst decay or activation during an experiment. Such activity changes can be quantified in some cases using temperature scanning methods, and these will be considered later.

Most of these assumptions are also inherent in isothermal experimentation but are usually hard to verify, often ignored, and generally under-appreciated. As we will see later, TS methods allow for ready verification of these assumptions before the definitive set of kinetic data is collected. That capability alone should improve the quality and utility of kinetic data when it is produced using temperature scanning methods.

The most important point is that in the TS mode of operation there is no need to wait for a reaction to reach isothermal steady state. This is a large part of the reason why, using TS methods, kinetic data can be collected so much faster than in the conventional isothermal steady state mode of operation. We must also resign ourselves to the fact that TSR methods often yield primary data that cannot be interpreted by conventional methods of data handling. This is disconcerting, and therefore the question that needs to be examined first is: how are valid reaction rates to be calculated from the plentitude of seemingly uninterpretable TS data?

It turns out that, for the TS-BR, rate calculations can be done exactly as they were in isothermal BR studies. For the TS-PFR and TS-CSTR, a more detailed mathematical investigation is required to develop the correct procedure for rate calculations. In each case, once the correct mathematical procedure has been established, a natural physical interpretation of what is going on, involving the concept of *operating lines*, becomes obvious. To capitalize on this the use of operating lines will first be described. Understanding the nature of operating lines will help us develop an intuitive understanding of the unfamiliar procedures used in TS methods.

Operating a Temperature Scanning Reactor

The basic TS mode of operation involves experiments, each consisting of a number of *runs*. Each run consists of operating the reactor over a period of time while the temperature of the feed, and usually of the reactor surroundings, is varied in some way. During each run, frequent (continuous if possible) measurements of temperature and conversion are made *at the outlet conditions*, i.e. of product drawn off from a TS-BR, or of product exiting a TS-PFR or TS-CSTR. Rates will later be calculated from this raw data obtained at the exit conditions present at the moment of sampling.

It is immediately obvious that a TS-PFR experiment might well provide abundant raw data, but even if reaction rates are extracted from this, the data will be at a variety of temperature/conversion conditions, and with unknown temperature changes along the axis of the reactor. In a TS-PFR the data will not be generated under isothermal conditions, as is required for interpretation of data obtained from traditional isothermal PFR reactors. Therefore a TS-PFR experiment will not *automatically* give sets of rates measured at a fixed temperature. More elaborate data analysis methods will have to be used to calculate reaction rates and pursue model fitting and parameter estimation. This issue is examined in more detail in Chapters 7 and 8, where it will be shown that the abundance of available rate data easily overcomes any potential disadvantage of not obtaining isothermal data directly. When one is interested in isothermal rate data it can be obtained from TSR results by a procedure called *sieving*.

Application to Various Reactor Types.

Consider the reversible reaction $A \leftrightarrows B$. The methods to be described are by no means limited to such simple types of mechanism and other, more complicated reactions, can be analyzed in exactly the same way. We will use the following notation, with units as indicated:

t	=	(clock) time into a run (s)
V	=	volume of fluid in reactor, or "active reactor volume" (l)
f_i	=	volumetric flow rate at the inlet (l/s)
N	=	number of moles of A in the reactor (mol)
X	=	molar fractional conversion of reactant A
C_A	=	concentration of reactant A (mol/l)
C_{A0}	=	initial concentration of reactant A (mol/l)
T	=	temperature of reactant
T_R	=	temperature of reactor body
T_C	=	temperature of catalyst
T_E	=	temperature external to reactor body
T_e	=	temperature at the exit of the reactor
T_I	=	temperature at the inlet of the reactor
δ_V	=	volumetric dilation factor (due to heating and reaction),
W	=	weight of catalyst in the active reactor volume (g)
$r(X,T)$	=	r_A = reaction rate = $(1/V)\, dN_A/dt$ (mol/s/l)

Temperature Scanning Batch Reactor

The traditional mode of operating a batch reactor requires attaining and maintaining some desired isothermal condition during the part of each run when data is being collected. The period during which the temperature is being stabilized affords no useful data, despite the fact that the reaction is proceeding, and information is being generated during that time. By contrast, in the TS-BR mode of operation, one deliberately allows the temperature to vary during a run, either by using internal or external heaters, and/or by allowing the heat of reaction to drive temperature changes. An investigator may choose to control the heating so as to drive the temperature through pre-selected conditions, but this will not affect the calculation of reaction rates.

The benefit of such a procedure would lie only in producing reaction rate data in some desired region of the temperature vs. conversion reaction phase plane. We note in particular that the data lost during the heat-up in an isothermal BR to obtain rate data at high temperatures are all kept for interpretation in the TS-BR. The TS-BR yields usable data at all times as long as the temperature of the reactive volume is kept uniform and recorded. We must only ensure that stirring within the reactor is vigorous and that the measured temperature correctly represents the bulk temperature of the reactant. We must also measure the composition of the reactor contents at frequent intervals. Since the reaction progresses during the time of the analysis there is a premium on devising a rapid, preferably continuous, measurement of composition in the reactive volume of the TS-BR.

Thus, as in the isothermal mode of operation, appropriate quantities are monitored continuously or at short time intervals in the TS-BR, and concentrations and temperatures are recorded as raw data. Since there is no volume expansion in a constant volume BR, the relationship between conversion and concentration is simply $X = 1 - C_A / C_{A0}$. Then the following rate equation holds for a constant-volume BR:

$$r_A = \frac{1}{V}\frac{d(C_A V)}{dt} = -C_{A0}\frac{dX}{dt} \tag{5.1}$$

In a constant pressure BR, or indeed any batch reactor where the volume may change by some factor δ_v, the rate equation becomes

$$r_A = -\frac{C_{A0}}{\delta_v}\frac{dX}{dt} \tag{5.2}$$

To see this, recall that the fundamental rate expression is

$$r = \frac{1}{V}\frac{dN}{dt} \tag{5.3}$$

where N is the number of moles of A in some quantity of fluid of volume V. Thus

$$r = \frac{1}{\delta_v V_0}\frac{dN}{dt} \tag{5.4}$$

where V_0 is the original volume of the fluid. Also, conversion is calculated from

$$X = 1 - \frac{N}{N_0} \tag{5.5}$$

where N_0 is the original number of moles of A in the fluid, so that

$$N = N_0(1-X) \tag{5.6}$$

Thus

$$\frac{dN}{dt} = -N_0 \frac{dX}{dt} \tag{5.7}$$

and so

$$r = \frac{1}{\delta_v V_0}\frac{dN}{dt} = -\frac{N_0}{\delta_v V_0}\frac{dX}{dt} = -\frac{C_{A0}}{\delta_v}\frac{dX}{dt} \tag{5.8}$$

as in equation 5.2.

In addition to the rate equation, there are the following heat transfer equations:

$$\frac{dT}{dt} = k_1\, r(X,T) - k_2(T - T_R) - k_3(T - T_C) \tag{5.9}$$

$$\frac{dT_R}{dt} = k_4(T - T_R) - k_5(T_R - T_E) \tag{5.10}$$

$$\frac{dT_C}{dt} = k_6(T - T_C) \tag{5.11}$$

where k_1,..., k_6 are various combinations of constants involving heat capacities, surface areas, heat transfer coefficients, etc. and could be developed in detail if required.

In practice there may be heat profiles within the reactor walls, so a complete set of heat transfer equations might be more complicated than equations 5.9 to 5.11, and some of the constants referred to may vary with temperature. Nevertheless, the main point applies: no matter how the system evolves, and no matter what more-complete heat transfer equations might apply, equation 5.1 holds at every instant. Thus, rates of reaction can be calculated by numerically differentiating the observed concentration C_A with respect to the clock time t. The caveat is that temperature changes between readings of the concentration must be small so that the rate calculated can be assigned to a well defined temperature. This requirement puts a premium on rapid methods of analysis in the operation of TS-BRs.

We emphasize that equations 5.9 – 5.11 do not have to be solved. Indeed, they are only suggestive of what a suitable set of equations might be. Furthermore, any suitably complete set of equations would surely be intractable. Our purpose in examining these equations is to see that rate-values can be measured by measuring dX/dt, directly or indirectly. This, of course, is quite obvious in the case of the batch reactor. We will see subsequently that it turns out to be the case for other reactors, where it is not nearly so obvious how to measure dX/dt. Measuring reaction rates using a TS-BR may therefore be simple in principle, but might initially confuse those accustomed to conventional methods.

Once the methodology is understood it becomes clear that the TS mode of operation is always considerably more productive than the isothermal mode. To begin, we do not need to wait for steady state or discard all readings before an assigned temperature is reached. Moreover, there is a great deal of freedom in how the temperature may be ramped. The operator will need to use different ramping rates and perhaps even very different ramping trajectories in the various runs of a TS-BR experiment drive the system into a wide range of X-T conditions during the several rampings (runs) that will constitute the experiment. The only constraint is that the maximum ramping rate does not exceed the capabilities of the analytical system, which should supply about one analysis per degree of temperature ramp.

From the data gathered in each run, one can then calculate reaction rates along the temperature trajectory of the run. This gives us a set of X-r-T triplets over a region of the X-T plane. Several runs, along different temperature trajectories, constitute a TS-BR experiment and fill out a wider region of the reaction phase plane. Sieving (and, where necessary, interpolating) this data can be made to yield sets of rates at various isothermal conditions. These isothermal rates can then be fitted to rate models in the usual way. Or, given the plethora of rate data available from such a TS-BR experiment, one can use more general fitting methods, detailed in Chapters 7 and 8, that directly incorporate the temperature variations.

Clearly, the temperature ramping policy that is optimum for purposes of sieving will be one that covers the area of the reaction phase plane with an evenly-spaced grid of points, so that the rates are well distributed over the range of conditions being investigated. The policy that is convenient and comes close to doing this involves linear ramping of temperature at ramping rates chosen from a specific series. This series consists of ramping rates exponentially distributed between the upper and lower ramping rates one intends to use. Figure 5.1 shows the conversion that might be observed in a reversible exothermic reaction if ramping rates between 1 and 40 K/min are used and the duration of the ramp is adjusted in each run so that the run is terminated some distance after maximum conversion is reached.

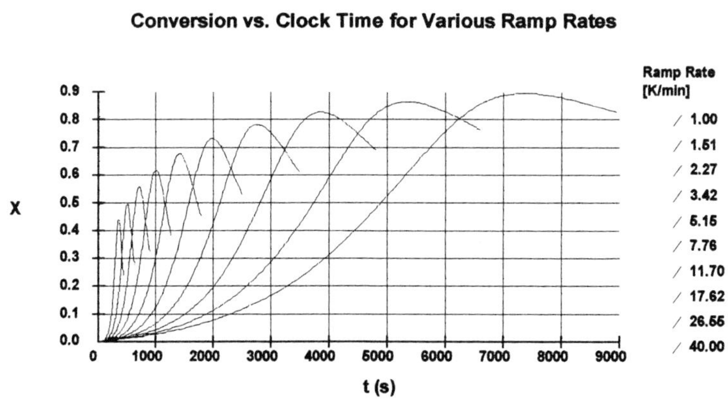

Figure 5.1
Conversion in a TS-BR as a function of clock time at various ramping rates. The slowest ramping rate of 1 K/min yields the flattest curve on the right. Successively steeper curves to the left correspond to ramping rates listed in sequence on the right of the figure.

Such a wide range of ramping rates needs to be possible in the TS-BR if a wide range of conversions is to be investigated over a broad range of temperatures. Note also that the runs vary greatly in duration due to differences in temperature ramping rate. The analytical system must be capable of collecting enough data under all these ramping conditions. By re-plotting this data on the reaction phase plane we see the result presented in Figure 5.2.

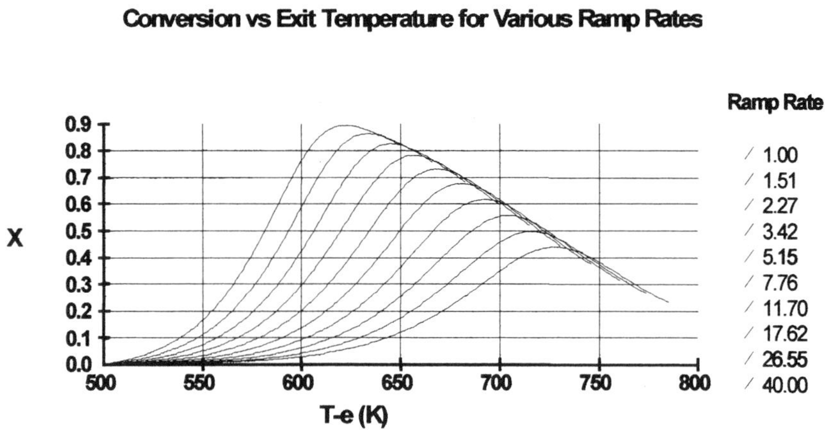

Figure 5.2
Conversion as a function of reaction temperature at various ramping rates. The slowest ramping rate of 1 K/min yields the steepest curve on the left. Successively flatter curves to the right correspond to ramping rates listed in sequence on the right of the figure.

Notice that the region of the reaction phase plane being investigated is covered by evenly spaced curves, making interpolations easy. This is the result of choosing an exponentially distributed set of ramping rates. The downward trend of the curves on the right is due to the back reaction of the reversible system in the case of exothermic reversible reactions. We see from Figure 5.2 that there is little information on the kinetics of the reaction in this late portion of the curves. All run conditions converge on the equilibrium curve. The point at which each run should have been terminated was therefore at the maximum conversion or shortly before. In practice the run would therefore be terminated at either a maximum permissible temperature or at the conversion maximum. In this simulation the temperature ramps extended from 500K to a maximum of 780K, as shown in Figure 5.3.

If the set of runs in a TS-BR experiment is appropriately spaced and sufficiently dense, triplets of conversion-rate-temperature can be obtained for any point within the enveloping curves on Figure 5.2. Conversion is obtained from interpolations on Figure 5.1; rates are obtained from slopes or numerical differentials of the same data. The corresponding temperatures are obtained for the ramping rate and clock time from Figure 5.3. An unlimited number of such triplets can be generated and sub-sets of isothermal or other constrained data can be selected, or *sieved* out, from this overall set for specific purposes.

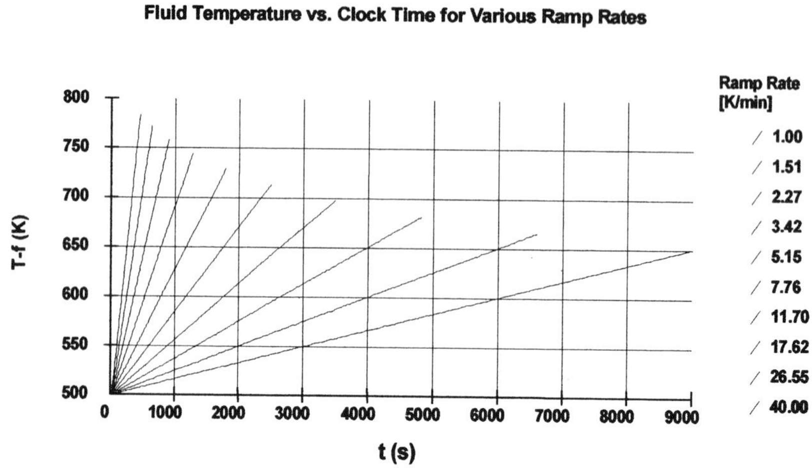

Figure 5.3
The slowest ramping rate of 1 K/min yields the curve with the lowest slope on the right. Successively steeper curves to the left correspond to ramping rates listed in sequence on the right of the figure.

Temperature Scanning Plug Flow Reactor

The process of getting reaction rates from temperature scanning plug flow reactor (TS-PFR) data is considerably more complicated than that for the TS-BR. For the TS-PFR it is not possible to obtain rates from a single run; instead, it is necessary to assemble data from several runs performed in specific ways. As we will see, the operating conditions of these runs (temperature ramping, space velocity, etc.) must be carefully controlled if we are to obtain valid rates from a TS-PFR. It turns out that the simplest suitable operating condition is when the space velocity is held constant during each run (at a different value for each run) and the temperature of the inlet feed is ramped in an identical manner for each run. Several such runs are required to complete a TS-PFR experiment. Other, more complicated operational procedures are also possible, and we will see later what advantages they might offer.

As it is with all TS techniques, rate data acquisition using a TS-PFR can be very fast and a vast amount of data can be obtained from a single experiment. Unfortunately, in the TS-PFR rates cannot be calculated in real time, as was the case with the TS-BR and will be the case with the TS-CSTR. Instead, rate data is obtained after all the readings from the several runs of a TS-PFR experiment are available. This post-experiment processing of the raw data will, however, produce the same X-r-T triplets as we discussed above. The triplets will be available over the whole range of X-T conditions covered by the experiment, just as they were for the TS-BR.

Operating Conditions for a TS-PFR

In order to collect meaningful data in a TS-PFR, the following basic operating conditions must be met:

1. The reactor must be of uniform effectiveness along its length (e.g. uniform diameter, uniform reactor body, and uniform catalyst packing).
2. The catalyst load and the reactor must be the same for each run.
3. Each run must start with the reactor in the same steady-state condition.

This starting condition need not be isothermal along the reactor but it is best to start with an isothermal profile and zero-conversion at the inlet. The simplest way of achieving this is to allow an *equilibration* time before each run, with the feed entering at a constant and low-enough temperature. The feed temperature and composition as well as the external temperature of the reactor should then be held constant until the system reaches a steady-state. We assume that the feed undergoes no conversion in the preheater, before it enters the reactor, under any of the experimental conditions encountered. Two more conditions are required.

4. The ramping protocol for the feed must be identical for each run. The ramping may be linear or non-linear, and may go up or down or both, so long as it is the same for each run.
5. The temperature ramping protocol of the external bath surrounding the reactor must be identical for each run. This ramping may be the same as that for the feed or different. Depending on the physical configuration of the reactor, it may be convenient not to ramp the bath temperature at all.

These constitute the *boundary conditions* for solving the governing differential equations that follow.

Mathematical Model of a TS-PFR

We begin with the simplest case by assuming that the feed rate is constant throughout each run. This is not an absolute requirement but, as will be seen below, any variations from this procedure must be analyzed with care and the data treated appropriately. As an example we will consider a typical experiment with about ten runs, each with a different but constant feed rate.

A critical issue that needs to be understood from the beginning is that in each run there are three different "times", which we will now define:

t = *clock time*, time into the run at the moment of input of a given plug of feed;
τ = *space time*, a defined quantity with units of time, discussed in Chapter 2;
s = *contact time*, length of time that a plug has spent in the reactor (Chapter 2).

In all kinetic treatment of PFR data, *space time* is used to define a nominal time, not necessarily the real time, that a plug spends in a reactor. Of the numerous definitions of space time possible, we will use the definition:

the reactor. These will then be used to calculate the corresponding conversions X_i via the relationship $X_i = 1 - \delta_v C_{Ai}/C_{A0}$. Calculation of expansion factors is discussed briefly below (at the end of CSTR section) and in detail in the first section of Chapter 7.

We now can make the simple but crucial observation that for any given fixed clock time t_i the evolution of the above equations tracks the X-T conditions of a specific plug, namely the plug that entered at time t_i. The boundary condition that results in the operating requirement that each run must start in the same steady-state condition (number 3 on the list) says, in effect, that each of the initial functions $X(\tau,0)$, $T(\tau,0)$, $T_R(\tau,0)$, $T_C(\tau,0)$ is the same for each run of an experiment. We have already noted that each of $T_I(t)$ and $T_E(t)$ is the same throughout each run. We also note that $T(0,t) = T_I(t)$. Finally, if we assume there is no conversion until the reactant enters the reactor, we have $X(0,t)=0$. In this way we have defined a full set of boundary conditions as listed above, conditions that completely determine the evolution of equations 5.18 to 5.21.

The boundary conditions guarantee that these equations will evolve identically for each run. This is a crucial point. Its recognition will allow us to assemble data from several different TS-PFR runs so as to construct the correct operating lines for the reaction/reactor combination under study, and thence to calculate the correct rates of reaction. Again, we emphasize that equations 5.17 – 5. 21 do not have to be solved. The techniques to be developed for measuring rates and fitting rate expressions do not require any particular solution to any particular set of reactor system equations.

The operating lines thus constructed "synthesize" the behaviour of a multi-port PFR reactor (see Figure 2.3 and associated discussion) as it would have operated with each successive input temperature. The several runs in the TS-PFR experiment offer us the conditions at succeeding ports along this multi-port reactor. Thus for a ten-run experiment we have data from ten such ports along each operating line. In principle we can trace an unlimited number of such operating lines from each experiment.

Such operating lines exist for all multi-port PFRs. There is no requirement that idealized conditions, such as adiabatic or isothermal, be present during reactor operation. Therefore the data from a TS-PFR need not be gathered under idealized operating conditions in order to construct valid operating lines from which rates can be calculated. This greatly simplifies the temperature control requirements in TSR operation and makes the TS-PFR relatively simple to build.

The procedure used in interpreting TS-PFR data is analogous to, and includes, that currently used to synthesize the behaviour of a conventional isothermal reactor. It is therefore equally valid. In conventional studies we synthesize the isothermal operating line by taking readings at the outlet of a fixed length reactor at various space velocities. These data are understood to correspond to measurements at various ports of a multi-port reactor operating at isothermal conditions. For this operating line we make a plot of X vs. τ and measure rates by taking slopes along this curve. Exactly the same procedure yields the large numbers of operating lines made available from each TS-PFR experiment.

We can sum up the concept of temperature scanning in a TS-PFR as follows: the method is based on the realization that operating lines for any reaction, under any thermal conditions, can be synthesized by collecting appropriate data under prescribed experimental conditions. The collected data involves measurements, at each point along the operating line, of the following:

- inlet temperature,
- outlet temperature,
- oven temperature,
- outlet pressure,
- output composition, and
- space time.

Processing this data involves several steps. The following list shows that several of these steps yield important new information and offer major improvements over conventional methods of data processing:

- error removal from raw data,
- mapping of data in various sets of coordinates in 2D and 3D,
- calculation of rates,
- sieving of data for various sets of interest, e.g. isothermal rates,
- fitting large numbers of data points to candidate rate expressions.

We note again that equations 5.18 to 5.21 present a greatly simplified model of the reactor. In particular, the constants $k_1,...,k_6$ (involving the heat of reaction, heat capacities, surface areas, heat transfer coefficients, and density of feed) may in fact vary with temperature and conversion. This does not affect the main point, namely that equations 5.18 to 5.21 are driven solely by the values of X and T and the other related temperatures, and by the boundary conditions, which are the same for each run. Hence the system of equations will evolve identically for each run. This holds true even if some of the k_i happen to be functions of X and T.

The constants k_i have various units, but they are all measured "per unit length of reactor", and so do not depend on the total length L of the reactor. We also make the assumption that the k_i are not strongly dependent on the linear velocity of the reactants and that the flow is turbulent, so that we have ideal radial mixing within each plug as it transits the reactor. The consequence of these assumptions is another key feature of the TS-PFR: the evolution of the system does not depend on the length of the reactor or on the linear velocity of the feed. The X-T conditions at any point in the reactor depend only on the space time τ corresponding to that run and on the clock time t at which that plug entered the reactor. For any given (τ,t) pair, the $X(\tau,t)$, $T(\tau,t)$ conditions of the plug will be identical for each run.

Interpretation of the mathematical model

In an actual TS-PFR experiment there will be several runs, each with a different constant feed rate. The model above shows that in a certain sense all the runs are identical. At any given clock-time t, each run will have a identical X-τ profile along the length of the reactor, and an identical T-τ profile. The temperature will therefore vary in some fashion along the axial direction. The only difference is that the profiles of the runs with shorter total space times (corresponding to their higher feed rates) will match only the first part of the profiles of the runs with longer total space times. All the profiles are identical except that some extend to higher τ-value than others. This is exactly what happens in a hypothetical multi-port reactor with the same axial temperature profile. It is also exactly the assumption made in running a conventional PFR. There we change space velocity from run to run and assume that by doing this in an isothermal, fixed-length reactor we are visiting different ports in a long isothermal multi-port reactor.

Having considered all these issues, we proceed to perform a simulated TS-PFR experiment. Figure 5.4 shows the results of a TS-PFR experiment that correspond to the raw data shown in Figure 5.1 for the TS-BR.

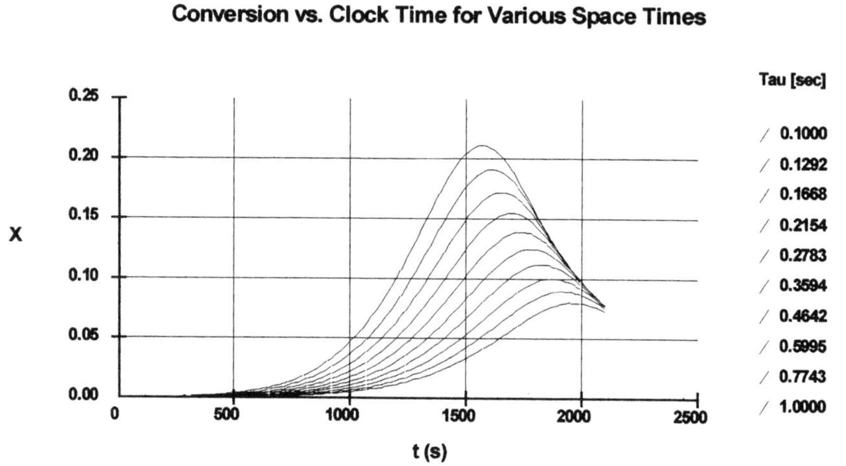

Figure 5.4
Conversion in a TS-PFR as a function of clock time at various feed rates. The fastest flow rate (shortest tau) of 0.1 sec yields the shallowest curve on the right. Successively steeper curves to the left correspond to longer space times listed in sequence to the right of the figure.

We can think of the data measured at any given clock time in a TS-PFR experiment as reporting the parameters of a single hypothetical plug of feed that enters the corresponding multi-port reactor at this clock time t_0 (and therefore at inlet temperature T_{I0}) and finds itself at successive ports (at different values of τ) along the reactor. To see this, let us perform the following "thought experiment".

As a plug travels along a multi-port version of the corresponding reactor it achieves a continuum of X-τ and T-τ values. Whenever this hypothetical plug reaches a τ -value that equals the total space time for one of the actual TS-PFR runs, its X-τ, and T-τ values are measured and made available as the corresponding values from the TS-PFR experiment. Each entering plug traces out the operating line for this reactor at an entrance temperature T_{I0}. However, we do not trace the fate of any given plug in a continuous way. Instead we catch snapshots of its fate in the various runs of an experiment. In each run we identify the plug of interest as the one that entered the reactor at a given clock time (and therefore inlet temperature). From the several runs of the TS-PFR experiment we therefore obtain several snapshots of the state of the plug as it traverses the synthesized multiport reactor, and thence construct the operating line.

During temperature ramping, successive plugs will trace out operating lines starting from successive different initial temperatures. Each of the operating lines, after smoothing, offers an unlimited set of rate measurements (slopes), just as it would for an isothermal reactor data set, as long as there are enough points along the line for reliable interpolation of the X-τ curve.

However, in a TS-PFR each rate at a given value of τ from any one operating line is normally at a different *output* temperature than it is at any other point from the same operating line. This leads to the absence of isothermal rates on any given operating line, a difficulty that is more than compensated by the fact that we have an unlimited set of such operating lines, each at a different *inlet* temperature. This allows us to interpolate and sieve-out a conversion and rate at a desired constant *output* temperature from as many of the available operating lines as we wish. The consequence is that the TS-PFR experiment yields an unlimited set of (X, r, T) data, each at a different value of τ, for each of the temperatures observed at the output. In each such set of *isothermal data* these results are available over the range of conversions that was observed at that exit temperature.

An example of a TS-PFR experiment showing ten runs and a selection of operating lines pertinent to that experiment and extracted from the raw data is shown in Figure 5.5. Each of the operating lines was constructed by selecting points with the same input clock time (and temperature T_{I0}) from the several runs of the experiment.

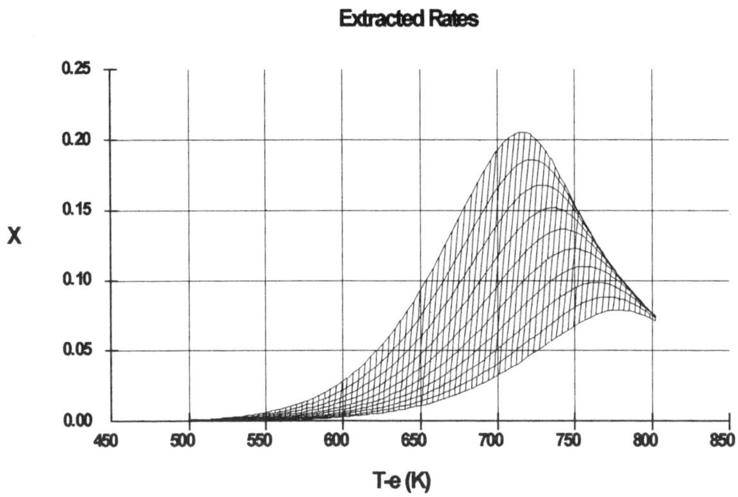

Figure 5.5
Selected operating lines sloping to the right and filling out the space of reaction conditions made available by a TS-PFR experiment. Every point within the area containing the operating lines can be made available for use in rate equation fitting.

The area lying between the highest and lowest curves contains points where all (X, r, T) triplets are available from this one TS-PFR experiment. The triplets, as noted, are the data required for fitting to mechanistic rate expressions. Points lying outside this area can never be made available using this reactor. Some of the outlying area may be accessible using a TS-PFR with a different L/D ratio or using a TS-CSTR, but they cannot be accessed by the reactor used here. This little-appreciated fact constitutes a limitation which is not specific to the TS-PFR. The same outlying points are not available from a conventional isothermal PFR with similar operating characteristics.

Operating Lines

The above discussion shows the importance of the concept of operating lines in understanding temperature scanning methods. They offer a useful and intuitive way to view X-T-τ data by envisioning the performance of a multi-port version of a long plug flow reactor, in particular the curve of X vs. T measurements along the reactor, presented in the X-T phase plane. In the case of conventional isothermal steady state operation this line is at a temperature that remains constant with clock time and contains the hidden dimension τ which increases along its length starting from the entry condition (see Figure 2.4).

The situation is somewhat different for a non-steady state TS reactor, and specifically for the TS-PFR. In that case each plug is moving along its own operating line, starting at the entrance condition for that plug. The various plugs present in the reactor at a given clock time are all on different operating lines. Their individual operating lines are then assembled from the several runs of a TS-PFR experiment by the application of the theory described in the defining equations cited above. This does not require us to solve the equations. We can assemble the operating lines using the simple procedure of picking points corresponding to the same clock time in each run of the experiment. These points present snapshots of the instantaneous condition at several ports, in a synthesized plug flow reactor. The ports are spaced at several axial positions (space times) along this synthesized reactor. They show a succession of fates that a plug of reactant would encounter as it proceeded along this particular imaginary reactor. This information can be assembled to yield operating lines for any of the plugs entering the reactor at successive clock times t_i.

Thus the operating lines of a TS-PFR are not constant with clock time but evolve with clock time as the inlet temperature changes. One way to visualize this is that the operating lines derived by operating the TS-PFR sweep out the shaded area in Figure 5.5 from left to right. The operating lines shown there are merely a few samples of the many positions of the operating line during that sweep. Each of these operating lines, allowing for interpolation, contains an unlimited number of rate data triplets. And there is an unlimited number of such operating lines.

In contrast, the operating line from isothermal operation will be a vertical straight line made up of several discrete points, collected in separate experiments, representing several space times located along the operating line. It would take as many isothermal experiments as there are runs in a single TS-PFR experiment to delineate this one operating line in comparable detail. Even then the error associated with the isothermal runs is more difficult to eliminate than it is in TS-PFR experiments. Then, it would take a great many such isothermal operating lines to define an equivalent area of the reaction phase plane in comparable detail. This is the reason for the large volumes of data available from TSR and why comparatively few points are yielded by conventional experimentation.

A series of steady state adiabatic reactor experiments, at different space velocities and the same inlet temperature, will also yield but a single operating line. In the case of an adiabatic reactor, where all temperature changes are due entirely to the heat of reaction, the operating line will be a sloping line. An exothermic reaction will yield a line with a positive slope, similar to the lines on Figure 5.5. As was the case with the isothermal reactor, it would take a great many adiabatic operating lines to sweep out an area equivalent to that on Figure 5.5.

For a non-isothermal non-adiabatic reactor (as is usually the case for a TS experiment) the operating line is still a well-defined and unambiguous curve of X vs. T. Now the operating line need not be a straight line at all. Its exact shape will depend in some complicated way on the heat released by the reaction and by heat transfers into and out of the reactor. However, the operating line still conveys the same information as in the classic cases: it provides a picture of the fate of a plug transiting a multiport reactor, with each point on the operating line giving the X-T conditions corresponding to a specific space time.

> *The theory of the TSR and its prescribed operating methods release us from the rigors of isothermal or adiabatic operation by revealing the position of the operating lines recorded in raw TSR data, regardless of the thermal regime pertaining to the given reactor/reaction system.*

Notice that this means that the TS-PFR can be operated under isothermal or adiabatic conditions as well as any other. One way to envision this flexibility is to see isothermal reactors as operating under conditions where the heat transfer is infinite, allowing the reaction to track the control temperature perfectly. Adiabatic reactors in that view have zero heat transfer and no heat is lost from the reaction. Temperature scanning reactors operate with any value of heat transfer coefficient, including the above two extremes.

> *The idea of classifying reactors by their heat transfer characteristics generalizes the view of reactor operations and puts temperature scanning in perspective. The equations describing temperature scanning are not heat-transfer-regime-specific, but present the general formulation describing reactor operation under all heat transfer conditions.*

From the many TS-PFR operating lines we can sieve out isothermal, or isokinetic, or any other restricted collections of data, for further analysis. TSR *kinetic* data is therefore not directly related to the available *raw* data.

In the paragraphs below we will see how data from a TS-PFR experiment is used to construct operating lines, and how from these one can calculate rates of reaction. It is the use of the operating lines that spares us from solving the defining equations of the TSR. This simplification makes interpretation of raw TS-PFR data relatively straightforward.

Figure 5.6 shows a set of unfamiliar, possible, and useable operating lines from a TS-PFR reactor operating in conditions where "overcooling" is taking place along the reactor. Along most of the operating lines, temperature is dropping despite the fact that an exothermic reaction is being considered. Nevertheless the results are valid and can be interpreted in the same way as those in Figure 5.5.

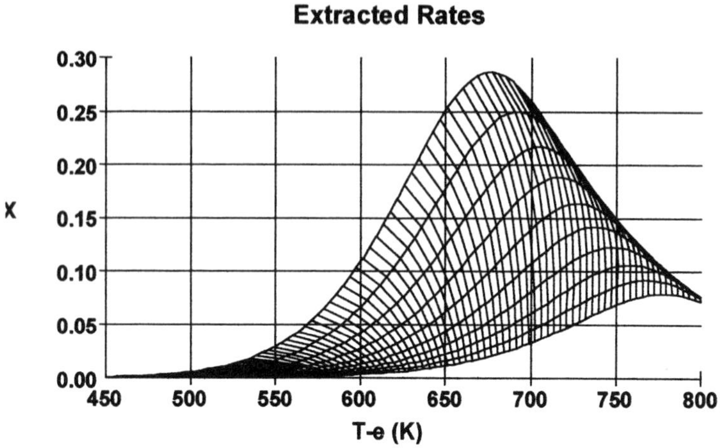

Figure 5.6
Operating lines for an overcooled reactor operation.

Calculation of Rates in a TS-PFR

In the discussion above we have seen how operating lines are formed from TS-PFR data in practice. Now we will examine why the procedure is valid and justify the claim that it is essentially free of assumptions and approximations. We will also examine how this procedure allows us to calculate rates of reaction.

Consider a typical TS-PFR experiment with ten (or more) runs at different feed rates. Let τ_i, i = 1,...10, be the total space time for the i'th run. For each run, we monitor the X-T_e conditions at the outlet of the reactor and hence obtain a large number of (X, T_e, t) triplets.

Thus for the experiment as a whole (i.e. for the ten runs) we obtain a large number of (X, T_e, t, T_I, τ_i) quintuplets. From this we can sieve out the data for any fixed t (hence fixed T_I), and find ten (X, T_e, τ_i) triplets. We can now plot X vs. T_e using this set of data. The points lie on the operating line *for this plug only*. We now draw an interpolated (not fitted) curve through these points to get a reasonable approximation of the corresponding operating line. A similar treatment of the X-τ and T_e-τ data will yield smooth curves in these co-ordinates. Finally, at any point on the smooth X-τ curve we can differentiate to calculate the local rate $r = C_{A0}\, dX/d\tau$ while at the same τ we read off the corresponding temperature T_e from the T_e-τ profile.

The rates from a single operating line are therefore calculated at whatever temperature happens to hold at the corresponding point on the T_e-τ curve. However, there are many such operating lines. If one wishes to get a collection of isothermal rates at some fixed temperature T_{ej}, then one needs to calculate many rates from many operating lines. Interpolation and sieving of this broader range of data is then made to yield an isothermal set of rates.

The procedure is as follows: we search along the T_e-τ curves of several operating lines to find on each the point where $T_e = T_{ej}$, the desired temperature. The τ value in each case is noted - say τ_k. Next, we search along the corresponding X-τ curves to find the point where $\tau = \tau_k$. The derivative dX/d τ taken at that point on each X-τ curve gives a rate r_k at the desired constant temperature. Repeating this procedure at a selection of values of T_e will produce sets of isothermal (X, r_k, T_{ek}) triplets suitable for isothermal data fitting. Such sieving procedures are ideally suited to data handling by computer and would be onerous beyond reason if done by hand.

The procedures described above will work well in the ideal situation, where there is no noise in the data and the X-τ curves represent the true conversion profiles so that the derivatives of the curves reveal the true rates. In practice there will be some noise in the raw data, and numerical differentiation notoriously amplifies noise, so that a straightforward application of these procedures is not likely to give satisfactory results. Instead, sophisticated numerical smoothing procedures must be used. The smoothing precedes the procedures presented above and leads to much more stable and accurate results by numerically filtering the data to remove experimental noise. The required filtering procedures are detailed in Chapter 7.

Comparison of Rates from a TS-BR and a TS-PFR

A comparison of equations 5.2 and 5.14 shows that exactly the same values for rate should result whether one uses the TS-BR or the TS-PFR. In the rate expression for the TS-PFR, "s" denotes the actual contact time, so that dX/ds corresponds exactly to dX/dt in the rate expression for the TS-BR. We never measure s; instead, we measure τ or t. Since the definition of t is straightforward, this brings us back to the question: exactly how should space time τ be measured in the TS-PFR? It is critical that we have this quantity clearly defined since τ is the time that is used in rate calculations and the issue of compatibility, indeed of the validity of the measures rates, hinges on this definition.

As defined above, space time is $\tau = V/ f_I$, where V is the reactor volume and f_I is the inlet volumetric feed rate. The unresolved question is: what is the reactor volume and what inlet feed rate is to be used? The two reasonable ways to measure inlet feed rate are:

1. the inlet volumetric feed rate at standard temperature and pressure (STP);
2. the volumetric feed rate of the fluid as it enters the reactor at the measured inlet temperature and pressure.

The advantage of the first is that τ then relates to a consistent measure of mass-flow rate that can conveniently be measured by instruments, and will have the same value throughout each TSR run. However, the actual inlet conditions usually differ greatly from STP, and the difference changes as temperature is ramped, so that τ measured in this way will differ greatly from the actual residence time s in the reactor. The advantage of the second is therefore that it uses the actual feed rate at entrance conditions. Its disadvantage is that, because the entrance conditions will vary during a TSR run, the measurement of space time will also vary during a run (although in a calculable way). Which measurement of space time is the correct one, in the sense of the one leading to the same rate values as those that would be obtained from the TS-BR?

The happy fact is that it does not matter which definition is used. In the key equation 5.11 that gives rates for the TS-PFR, namely $r = -C_0dX/d\tau$, both C_0 and τ depend on how one chooses to make the "initial" measurements. If one measures at STP, one gets certain values for C_0 and τ, and if one measures at actual entrance conditions one gets other values that both differ from the STP values by the same factor, namely by $1/\delta_0$, where δ_0 is the expansion factor from STP to inlet conditions. Thus the ratio C_0/τ is the same in either case, and so the rate $r = -C_0dX/d\tau$ will be the same since $dX/d\tau$ will also change by the same factor. Of course, when one converts between fractional conversion X and concentration C via the formula $X = 1 - \delta_vC/C_0$, one must use the C_0 at whatever initial conditions are being used, and δ_v must be measured from those same initial conditions. Again, it does not matter what initial conditions are chosen; the ratio δ_v/C_0 will be the same for all choices.

The measurement of V is less well defined, especially in the case of catalytic reactions. This issue was discussed in Chapter 2, and the final conclusion is that, if the effective volume V is defined in the same way in both the BR and the PFR, comparable rates will be measured in the two configurations.

Temperature Scanning Continuously Stirred Tank Reactor

The operation and description of a temperature scanning continuously stirred tank reactor (TS-CSTR) is, in principle, much simpler than for the TS-PFR. It turns out that rates can be calculated from each individual point in each run, and that flow rates and temperature ramping do not need the same careful control as the TS-PFR. Nevertheless, the operation of the reactor should approach the perfectly mixed condition very closely. Although in practice it may be difficult to make the necessary physical arrangements for complete and instantaneous mixing within the reactor, as with other TS reactor types there are verification procedures that will reveal if proper operating conditions are not being met.

Mathematical Model of the TS-CSTR

The variables C_A, C_{A0}, X, T, T_I, T_E, T_R, and T_C all have the same meaning as before, and once again they are all functions of clock time. Note that the variables C_A, X and T refer to internal conditions in the reactor, but because of the assumption of perfect mixing they are identical with those that can be measured at the outlet. In addition, we denote:

δ_V = volumetric dilation factor (due to heating and reaction),
f_0, f = inlet and outlet flow rates,
τ = space time = V/f_0,
N = moles of A in the reactor at any time t.

As in the TS-PFR, volume expansion plays an important role. It is related to inlet and outlet flow rates in a rather more complicated way than might be supposed. If the reactor were in steady state then we would simply have $f = \delta_vf_0$. However, in the non-steady-state case of the TS-CSTR, δ_v itself may be changing over time, and this causes an additional effect on the outlet flow rate.

Consider a short clock-time interval dt. In this time interval a small amount of feed, of volume f_0 dt, enters the reactor, expands by the factor δ_v to volume $\delta_v f_0$ dt, and forces an equal volume out of the reactor. In addition, during the same time interval the volume expansion factor pertaining to the contents of the reactor increases by some amount $d\delta_v$, and this causes the whole body of fluid in the reactor to expand from total volume V to $V(\delta_v + d\delta_v)/\delta_v$. The extra volume $Vd\delta_v/\delta_v$ thus generated is therefore also forced out of the reactor. Combining these two effects, we see that the volume exiting the reactor in time dt is $dV = \delta_v f_0\, dt + Vd(\delta_v/\delta_v)$. The outlet flow rate, f = dV/dt, is then:

$$f = \delta_v f_0 + \frac{V}{\delta_v}\frac{d\delta_v}{dt} \tag{5.22}$$

We can now get expressions for the flow rates of reactant A in and out of the reactor. The inlet flow rate of A is $C_{A0}\, f_0$. Also, since for a dilation factor δ_v the relation between conversion and concentration is given by:

$$C_A = \frac{C_{A0}}{\delta_v}(1-X) \tag{5.23}$$

it follows that the outlet flow rate for A is:

$$C_A f = C_A \delta_v f_0 + C_A \frac{V}{\delta_v}\frac{d\delta_v}{dt} = C_{A0} f_0 (1-X) + C_A \frac{V}{\delta_v}\frac{d\delta_v}{dt} \tag{5.24}$$

From the definition of rate r_A, we see that the rate of loss of moles of A in the reactor due to reaction is $-r_A V$. Putting all these rates together, we find that the net rate of accumulation of moles of A in the reactor is:

$$\begin{aligned} \frac{dN}{dt} &= \text{input rate} - \text{output rate} - \text{rate of loss by reaction} \\ &= C_{A0} f_0 - C_{A0} f_0 (1-X) - C_A V \frac{1}{\delta_v}\frac{d\delta_v}{dt} + r_A V \\ &= C_{A0} f_0 X - C_A V \frac{1}{\delta_v}\frac{d\delta_v}{dt} + r_A V \\ &= \frac{C_{A0} V}{\tau} - C_A V \frac{1}{\delta_v}\frac{d\delta_v}{dt} + r_A V \end{aligned} \tag{5.25}$$

We note that in the special case of steady-state operation, when dN/dt = 0 and $d\delta_v/dt = 0$, this reduces to the usual design equation for a steady-state CSTR:

$$-r_A = \frac{C_{A0} X}{\tau} \tag{5.26}$$

$$k_m = k_{m,in}\,(T_{out}/T_{in}) = (R/P_{in})(T_{out}/T_{in})$$

where P_{in} is the pressure at which the experiment is performed.

The effect of desorbed material on volume expansion will usually be small. It can readily be estimated from equation 5.66 by examining the value of k_mC_A. Since the desorption expansion factor refers only to the highly diluted content of A in the sweeping stream, it can generally be ignored if found to be close to 1 over the range of temperatures studied. Nonetheless, this type of expansion should also be incorporated in equations 5.57 and below, in equation 5.58, if accurate simulations are to be run.

A simple way of measuring desorption rates

There is yet another expansion factor we can define, the *I/O expansion factor*, which we take as the ratio of the outlet flow rate to the inlet flow rate, allowing for the expansion in total flow. From the discussion preceding equation 5.59 we see that the outlet flow rate, allowing for thermal expansion as well as desorption is $(T_{out}/T_{in})f_0 + k_mV_sr_n$ while the expanded flow of sweeping gas alone at the outlet will be $(T_{out}/T_{in})f_0$. Hence the I/O expansion factor is

$$\text{I/O expansion factor} = \Delta_{I/O} = 1 + k_mV_s\,r_n\,/((T_{out}/T_{in})f_0) \qquad (5.68)$$

We note that $\Delta_{I/O}$ depends on the current rate, r_n, of desorption, which we do not know, plus various known system constants. Since the other terms in equation 5.68 are known, one can in principle measure the net rate of desorption r_n using the I/O expansion factor alone, no analysis of the output stream is required and one is not troubled by the mixing going on in the void volume. When we include the expansion factor for k_m the temperature ratio for k_m cancels out that for f_0 leaving the very simple expression:

$$r_n = (\Delta_{I/O} - 1)f_0\,/k_{m,in}V_s \qquad (5.69)$$

The I/O expansion factor will not be much different from the molar expansion factor but it may be measurable. If so, it may provide a simple way of measuring net desorption rates. The kind of device that would make this technique viable is a "differential flow rate *or* differential flow velocity" meter. Since temperature does not enter into equation 5.69, this device would compare the flow rates of the inlet and a *cooled* outlet stream at a convenient common temperature, say the temperature at which inlet flow is metered.

A simple detector from a gas chromatograph could do this, were it not for the change in composition between the inlet and the outlet. Then again, a calibration of the GC detector will reveal the amount of A in the outlet. In that case, after correction for expansion in the outlet stream due to desorption, the I/O expansion can be calculated from the chromatographic analysis. Whether at this point there is any advantage to be gained over the direct application of equation 5.67, is not clear.

Calculating volumes from TS-CST-SSR data

The criterion that defines an equilibrated system is $r_n = 0$ and therefore

$$(1/\tau)(T_{out}/T_{in})C_A = -(V_v/V_s)\, dC_A/dt \qquad (5.70)$$

$$C_A = -(T_{in}/T_{out})(V_v/f_0)\, dC_A/dt \qquad (5.71)$$

This expression can be traced throughout the experiment and, if found to be true, allows us to calculate equilibrium constants at the conditions where it applies.

One can use a rearranged form of equation 5.71 to correct the often dubious estimate of what exactly is the effective void volume, V_v, in the reactor. The total volume of the reactor cavity V_t is easy to establish but the volume of the solid V_s is usually determined by other, off-line, methods. The restriction in all cases is that

$$V_v + V_s = V_t - V_i$$

where the volume, V_i, is that occupied by reactor internals.

Examination of Figure 5.16 shows that the net rate becomes essentially zero past about 350 on the clock time. After that, the system is near equilibrium and the net rate of desorption is near zero while the void volume continues to be flushed of accumulated material by the sweeping gas. A plot of V_v against clock time based on equation 5.71 will show that beyond the time where the rate descends to zero, the value of the volume V_v approaches its correct value asymptotically. In Figure 5.17 the volume correctly approaches 60 liters used in the simulation. Before that time the equation does not hold, equilibrium is not closely approached, and the trace of the equation tells us little about the system.

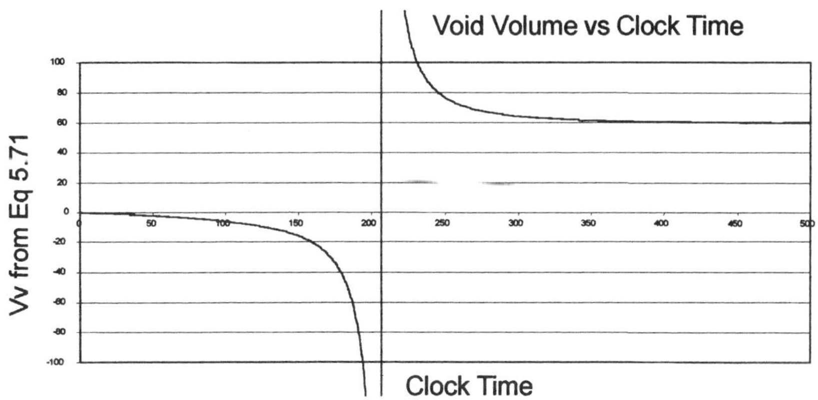

Figure 5.17
The right-hand branch of equation 5.17 is seen approaching the correct value of the void volume, V_v, asymptotically. The vertical asymptote identifies the clock time at which the maximum in output concentration occurred.

In cases where the reactor volume minus the internals can be measured, one can obtain the volume V_v from equation 5.71 and then

$$V_s = V_t - V_i - V_v$$

making the effective volume of the solid, V_s, also available from the same experiment.

The values of these two effective volumes constitute useful information for calculating the correct space time and obtaining a better understanding of the system, particularly the relation between the bulk volume of the solid as measured off-line and its effective volume in the reactor.

Calculating the surface coverage

Starting with a fully covered surface, the *correct* total amount of adsorbate released to the point where no more A is present in the output can be calculated by equation 5.56. The amount released up to any time t_0 can also be calculated from equation 5.56. The difference between the total and partial areas at times t_0 represents the number of moles remaining on the surface. This quantity, divided by the total moles released, is the fraction of surface covered, θ, at that time t_0 and therefore at that temperature on the ramp. Values of coverage obtained in this way may be used to calculate equilibrium constants.

Measuring Equilibrium Constants

A general procedure

There is a method that does not, strictly speaking, use TS procedures but can yield a series of discrete equilibrium constants using the TS-CST-SSR apparatus. The procedure is a *punctuated temperature scan* and is based on the following premise: if the apparatus is used as an isothermal batch reactor (that is, sweeping flow is stopped, and the sample is maintained at the temperature attained at that time), the gas phase will eventually equilibrate with the adsorbed phase. This is confirmed by equation 5.69 which states that when the flow rate f_0 goes to zero the net rate of desorption also goes to zero. When one observes no change in gas phase concentration with clock time after the flow and ramping is stopped, the system is at equilibrium and the concentration and temperature data can be used for calculating the equilibrium constant. Readings of this kind can be taken at several points during a ramp to ensure that equilibrium conditions are established. Volume expansion will not influence the results under the zero flow conditions proposed.

In the simplest case we can write the equilibrium between adsorption (a) and desorption (d) using equations 5.39 and 5.40.

$$k_d\,\theta = k_a\,(1-\theta)C_{Ae} \tag{5.72}$$

After rearrangement we have

$$K_e = k_a / k_d = \theta/(1-\theta)C_{Ae} \tag{5.73}$$

where K_e is the equilibrium constant. Since θ and $C_{Ae} = C_A$ and can be made available at each temperature where the scanning is paused, or at each temperature where the system was at equilibrium during scanning, θ can be calculated using equation 5.56 as described previously. From this one can calculate K_e at each pause.

A simplified procedure for evaluating K_e

An even simpler procedure for determining K_e uses just the equilibrated portion of a single ramping. At the time and conditions where the observed rate in Figure 5.15 and the calculated volume in Figure 5.17 reach their asymptotes (about 350 on the time scale used there), equilibrium is also approached. A plot of $\ln(K_e)$ from equation 5.73 versus 1/T plots the experimental K_e values on an Arrhenius plot. As long as the plot is curved, equilibrium is not near. This is true for times to the left of the asymptotic conditions in Figures 5.15 and 5.17. However, the portion of the curve that is almost linear at high temperatures corresponds to region where the system approaches the asymptote on Figures 5.15 and 5.17 and can be used to give an estimate of the Arrhenius constants for K_e over that temperature range.

If the range of temperatures is adequate, and measurement errors are small enough to establish the temperature coefficients of the equilibrium constant, then this single ramping can be used to calculate all of: the equilibrium constant, the desorption, and the adsorption rate constants for the system. In practice the best operating conditions for this type of run are at high values of *transit time*, i.e. under conditions where V_v/f_0 is large but there is a premium on accurate data to avoid the instabilities that result from the form of the equations used.

Notice that the right side of the curve can also produce a good straight line in some cases, potentially encouraging misinterpretation. The region where one should seek the straight line and the Arrhenius parameters describing the behaviour of the actual equilibrium should be first identified from appropriate regions in Figures 5.15 and 5.17.

A more reliable way to get an estimate of the Arrhenius parameters of the equilibrium constant is to take slopes of the trace in Figure 5.18 and plot these against clock time. If appropriate run conditions for estimating equilibrium constants were used, and the data is good enough, one will find that the resultant curve approaches an asymptote in activation energy at long ramp times (high ramp temperatures). Usually the curve will fail to reach the asymptote and one has to resort to various methods of estimating its value by extrapolation from the available data. This procedure will at the very least yield an upper limit to the value of the energy term of the equilibrium constant. A similar procedure will help to establish the asymptotic value of the frequency factor.

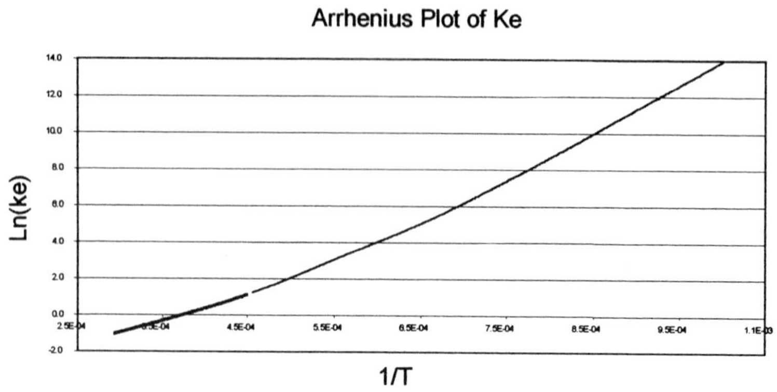

Figure 5.18
Plot of the equilibrium constant K_e from equation 5.73. The curve on the right is that corresponding to dis-equilibrated conditions. The portion to the left shows linear behaviour and corresponds to near-equilibrium conditions. The slope and intercept of this part of the curve give the frequency factor and activation energy of the equilibrium constant.

A General Method of Quantifying Rate Constants in Adsorption

If equilibrium constants can be obtained by the above methods the measurement of desorption rate kinetics is greatly simplified. An experiment is performed where the ramping rate is adjusted so that equation 5.70 is *not* obeyed over most of the temperature range examined. From this experiment one can calculate a large number of values of r_n at various temperatures.

By combining the net desorption rate, equation 5.41, with the expression for the equilibrium constant one gets:

$$\text{net rate of desorption} = r_n$$

$$= k_d\,\theta - k_a\,(1-\theta)C_A$$

$$= k_d\,\theta - K_e k_d(1-\theta)C_A \qquad (5.74)$$

We see that if K_e is known one can obtain the net rate of desorption, at any concentration C_A, in terms of one unknown, k_d in the equation above. The desorption rate constant is thereby made available without controlling the concentration of C_A in the output. This removes the need to keep the concentration of C_A low and allows for a much less complicated experimental procedure than that envisioned earlier. We no longer need to alter the flow rate in order to keep the C_A low to prevent re-adsorption. Re-adsorption is now a welcome aspect of the procedure.

The preferred way of studying adsorption/desorption equilibria and kinetics is therefore a one- or two-run procedure, as follows.

- Determine the equilibrium constant as a function of temperature from a "punctuated temperature scan" experiment or from just one ramping where equilibrium is established toward the end of the ramping.
- Determine the desorption rate constant as a function of temperature from a run at a ramping rate that does not allow equilibrium to be established, at least over part of the lower temperature range being investigated.
- From data obtained in the one or two experiments described above, calculate the adsorption rate constant as a function of temperature.

The simplest of these procedures will require just one experiment and will yield the void volume, the volume occupied by the solid, as well as the adsorption rate constant, desorption rate constant and the equilibrium constant as functions of temperature, without any concern about the level of C_A during the ramping. Whether one can implement the high transit time conditions, and reduce experimental error to the point where this "silver bullet" experiment can be carried out will depend on the hardware/kinetics combination of the system.

It may be surprising that all the rate parameters, together with their temperature dependencies, are available from just one or at most two runs. This is so in this case because of the simple and known expression for the kinetics of adsorption/desorption and because we have contrived to obtain the equilibrium constant, together with its temperature dependence, in one run. A full TSR experiment, with several runs, might still be necessary if the rate expression is not known *a priori* or if other issues, such as diffusion, need to be addressed. The methods used in such cases are described in the preceding discussion of the TS-PF-SSR.

General Observations Regarding TS-SSR Operation

The equations developed here allow the retrieval of large amounts of quantitative information on adsorption/desorption kinetics and thermodynamics from relatively simple experiments. The required apparatus is simple and variants of the configurations described already exist in many laboratories. The only sophistication required is in the precision of concentration measurements and in the control of run conditions.

The experimental methods and the quantification of data from a TS-CST-SSR is fully delineated by the above treatment and its trivial but interesting extension to cases where the initial concentration of adsorbent in the sweeping stream is not zero. The same equations can be used to quantify commonly available TPD results as long as the reactor configuration and run conditions conform to the assumptions used in the derivations presented here. Since most TPD experiments are carried out in plug flow configurations one can take V_v=0, but the volume expansion factors remain necessary if one intends to calculate the correct quantities desorbed from the sample and/or to quantify the kinetics and thermodynamics of adsorption/desorption.

Advanced Scanning Modes

Scanning Modes for the TS-BR

For the TS-BR there is no restriction on how the temperature is ramped. In exothermic or endothermic reactions the temperature will usually be driven mainly by the heat of reaction. However, if the reactor is constructed to allow for excellent heat control, the operator may attempt to control the temperature in such a way as to drive the reactant temperature in various specific directions. Even then, maintaining strict thermal control in highly exothermic reactions may be difficult. Trying to drive the internal temperature exactly along a prescribed profile will usually require more control capability than is affordable or necessary. Fortunately, for most purposes it should be sufficient to drive the internal temperature *approximately* along a desired profile, and this is always possible to do. The goal is simply to move the reactor temperature so that during the multiple runs of an experiment a broad area of the X-T plane is covered and trajectories are not repeated.

Unless the reaction is reversible there is usually no sense in ramping down after ramping up. On the other hand, ramping rapidly at first and then more slowly is frequently desirable. In this way the reactants are brought to a temperature where measurable reaction takes place and the ramp is then slowed down to allow for taking a larger number of samples at temperatures of interest. The salient differences between the procedures used in TS-BR operation and those used in standard methods of BR operation are that in a TS-BR samples are taken at all times during the ramps, the final condition need not be at some fixed temperature, and valid data can be extracted all along the way. In this way no time is wasted and more data is acquired.

Scanning Modes for the TS-CSTR

Similarly, for a TS-CSTR there is no restriction on the temperature ramping used, either of the inlet feed or of the external temperature. In this case the extra control allowed by freely varying the temperature of the inlet feed should make it much easier to guide the reaction to desired regions of the X-T plane. Note that it is not even necessary to have very strict control of the inlet temperature; it can vary rather freely without affecting the validity of the calculation of reaction rates. If the flow-rate method as in equation 5.37 or the β vector method described in Chapter 7 is used for calculating the dilation factor δ_V, it is not even necessary to measure the inlet temperature. All that is required is frequent measurements of the reaction (outlet) temperature and of the composition of the reactor contents. Figure 5.19 shows an experiment involving a set of ten arbitrarily designed TS-CSTR runs, each with a different "broken" (i.e. two stage) ramp.

Interestingly enough, valid reaction rates are available from all the ramps shown, and isothermal data sets are available over the whole range of temperatures by sieving the available data. In operating a CSTR the concept of individual ramps becomes moot. One can in practice meander over the reaction phase plane by ramping temperatures (and/or flow rates, as we will soon see) up and down without repeating any specific pattern. There is no need to terminate the meandering at any point before all the required data is assembled.

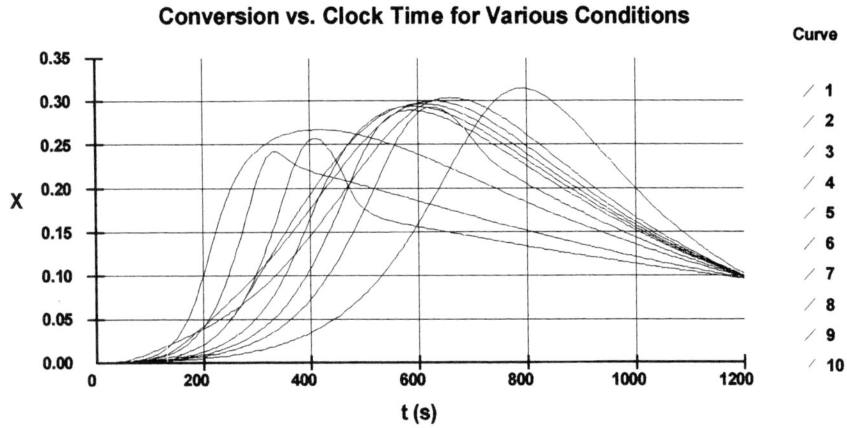

Figure 5.19
Conversion as a function of clock time using a set of arbitrary broken ramping rates. In these runs the ramping rate was kept linear for some time and then reset to a new linear ramping rate for the rest of the run. The breaks and the two segments of each ramping rate are not the same in each run. More than one break can be used in each ramp.

Scanning Modes for the TS-PFR

For the TS-PFR the situation is quite different. For the first run of each experiment the temperature ramping may indeed be chosen freely, and it need not be linear, or even always upwards. However, exactly the same ramping must then be used for the subsequent runs of that experiment. This clearly poses a significant control problem unless a very simple ramping scheme, such as a linear ramp, is used.

One might consider an up-down ramping scheme in which the temperature is ramped linearly upward for a time, and then ramped linearly downwards for a time. The goal would be to gather useful data during the time that would otherwise be wasted between runs waiting for the reactor to cool down. Unfortunately, experience indicates that the downward ramp often follows almost the same curve in the X-T plane as the upward ramp, so nothing much is gained.

An alternative procedure is to change the space velocity for the down ramp. Figure 5.14 shows such an up-down experiment where the left side corresponds to the up ramps of Figure 5.2 while the right side shows the down ramps at the same but negative ramping rate each down ramp at twice the space time of its preceding up ramp.

Although Figure 5.20 looks much different from Figure 5.2 and promises to yield additional data, it often does not. To see this we need to re-map the results on to the reaction phase plane where in Figure 5.21 it becomes clear that in this case the up and down results almost coincide. Notice how some of the up and down ramps tend to overlap or show up as closely spaced pairs, indicating that in these cases the data collected in an up ramp is very similar to that from a down ramp. The spread between the up and down ramps will vary somewhat with the degree of isothermality of the reactor but the main point remains unchanged; little is gained from this form of up-and-down ramping. The run corresponding to the shortest tau in the up-ramping set is the lowest. A careful variation of the flow velocities in the down ramping portion could improve the situation but the desirable flow velocities would require an extensive trial and error search to be identified.

However, a somewhat more elaborate up-and-down ramping scheme is fruitful. At the end of each upward run one must change the inlet flow rate and then let the system sit for a while to achieve steady state at this high temperature. Then the reactor is ramped downward, with data collected at the new flow rate. The downward ramps done this way obey all the required boundary conditions and therefore constitute a whole new experiment, with a whole new set of space times. The downward runs of this new experiment are now well interspersed between the upward runs of the original experiment. The X-T curves traced by runs of the new experiment will be roughly parallel to, but significantly offset from, the X-T curves of the original experiment. Their exact placing will depend, of course, on the space times and ramping rates chosen. Using this procedure, two experiments are done in roughly the time it takes to do one, if up-ramps are followed by unmonitored cooling periods.

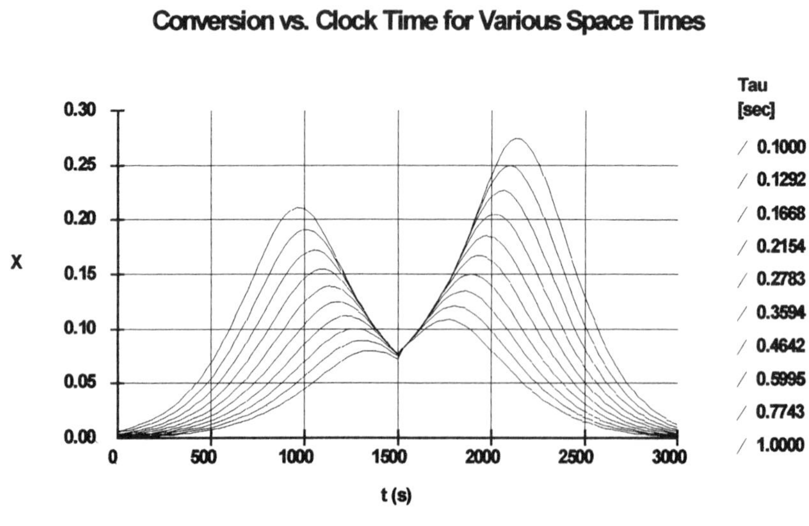

Figure 5.20
*Up-down ramping experiment in a TS-*PFR *where the left side corresponds to the up ramps of Figure 5.2. The right side shows the down ramps at the same but negative ramping rate. The lowest curve corresponds to the lowest tau (highest flow rate) listed on the right and each down ramp on the right is at twice the space time of its preceding up ramp. Because of the change in the space velocities in the down ramps, this data-set covers a broader range of conversions.*

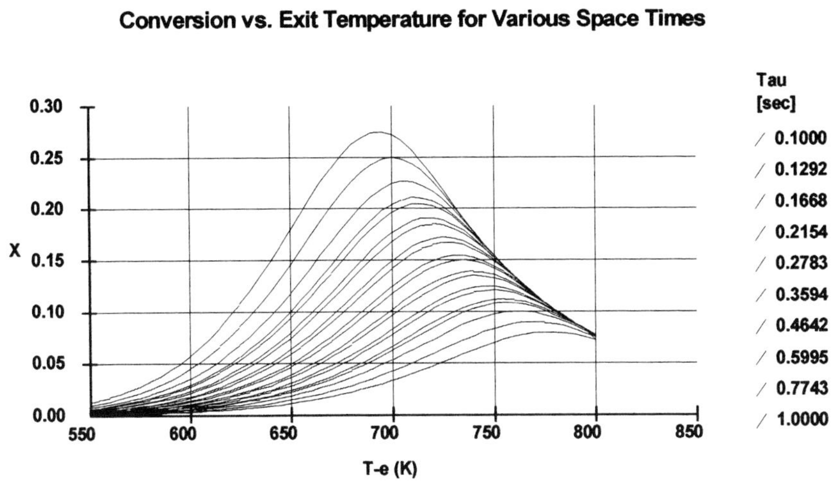

Figure 5.21
Here we see the results shown in Figure 14 replotted on the reaction phase plane.

Flow Scanning Modes

So far we have been considering temperature ramping only, with flow rates held constant. In this section we consider the possibility of varying the flow rate during a run. The TS-CSTR turns out to be very simple to deal with, with marvelous possibilities for interactive control. The TS-PFR, as usual, requires much more careful consideration. In both the plug flow and CSTR reactors, flow scanning can be used alone or in combination with temperature scanning.

Flow-Scanning in a TS-CSTR

Equations 5.13 to 5.18, used to calculate rates in the TS-CSTR, and equation 5.24, used to update the calculation of δ_V, all hold instantaneously whatever the current values of τ, X, δ_V, and dX/dt happen to be. Thus rate calculations for the TS-CSTR can proceed exactly as before, regardless of any variation in flow rates. It follows that the operator has complete freedom to ramp both the temperature and the flow rate of the TS-CSTR in any way that seems desirable.

One simple operating mode would be to imitate the up-down ramping mode described above for the TS-PFR: i.e. ramp the temperature linearly upwards, then change the flow rate, then ramp linearly downwards, all the while collecting X-T information. Doing this for several runs with different combinations of flow rates would allow good coverage of the X-T plane in a TS-CSTR experiment.

A much more interesting and interactive mode consists of a "joystick" controller, with motion in the X-direction controlling temperature ramping (up or down) and motion in the Y-direction controlling flow rate (higher or lower). By moving the joystick and observing the reactor output, the operator can move the reactor output to interesting regions of the X-T plane. The joystick controllers do not even need to be particularly precise, since the rate calculations are done on the basis of actual input/output measurements, not on the basis of controller settings. Figure 5.22 shows the raw data for a whimsical set of such runs.

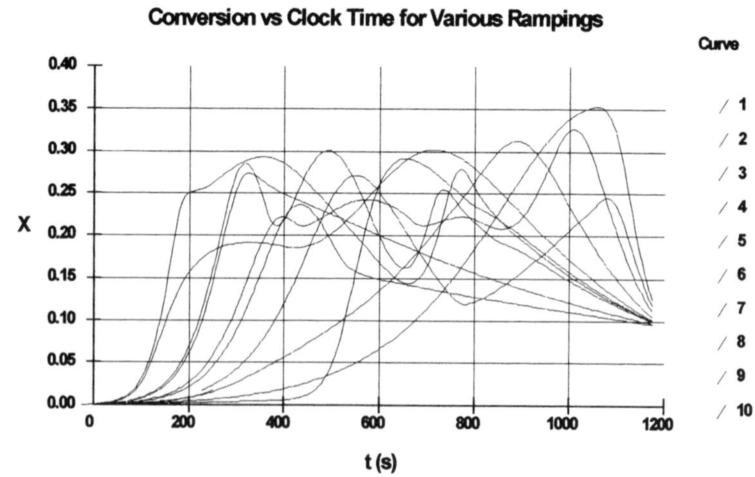

Figure 5.22
A set of conversion vs. clock time data obtained by changing both flow rate and temperature ramping in each of ten runs of a TS-CSTR. The temperature ramps used are those shown in Figure 5.23 while flow rate changes are shown in Figure 5.24

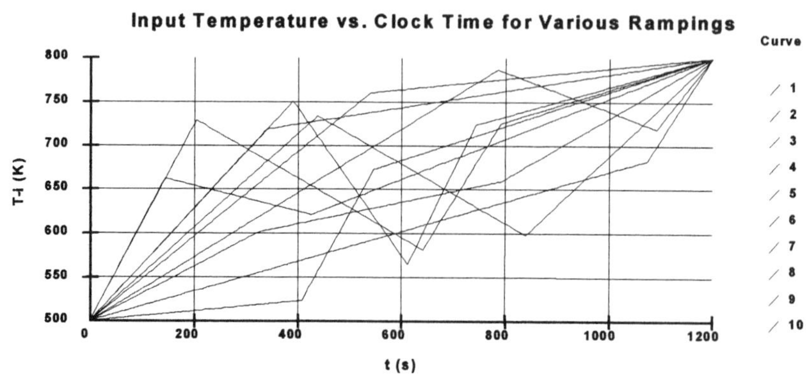

Figure 5.23
A set of arbitrary broken linear ramping schemes, one per run of a TS-CSTR whose output conversions are shown in Figure 5.22. A little thought will lead to the identification of specific temperature trajectories shown here with corresponding conversion trajectories in Figure 5.22.

In performing this experiment both flow rate and temperature were varied in each run. Figure 5.23 shows the changes in temperature. At the same time the flow rate was varied in the way shown in Figure 5.24.

It would be extremely difficult to predict by calculation what sequence of joystick movements would be needed to produce a desired X-T trajectory, and in fact impossible unless all the details of the reaction were known ahead of time. However, a human operator could quickly develop considerable intuitive expertise through repeated trial-and-error experience with a simulator or with a real reactor. This would raise the role of the technician in charge from the traditional "reactor operator" to a "reaction phase plane pilot" whose skill in covering the desired area of the reaction phase plane would be a great asset to expediting experimental work.

The trajectories used in this procedure are clearly not just simple ramps. They would in fact be quite complicated meanders through the X-T regions of interest. Figure 5.25 shows the meanders that result from the raw data shown in Figure 5.22. It is clear from Figure 5.25 that control of flow and temperature in tandem allows one to visit regions of the X-T plane that are inaccessible using temperature variations alone. The regions accessible by this simper but more restricted procedure are those that depart from the smooth up-down curves seen in Figure 5.25. It needs to be remembered that, during this seemingly disorganized meandering, valid kinetic data is being collected.

Once a desirable X-T trajectory has been established, by trial and error or otherwise, it may be useful to apply that trajectory to future standard tests in a development program for a particular type of catalyst. Such a procedure could test the catalyst at a number of X-T points of interest in one run. Although it is not feasible to calculate the required sequence of joystick movements, it is easy to record the movements that are actually used by the operator. This recording can then be used to automate all subsequent test runs for variants of the catalyst being developed. Standard tests of this type will not yield the volume of kinetic information required for fitting a rate equation but they promise to be much more informative and no more demanding than the single point catalyst evaluations currently done in development programs using steady state PFRs, a method of evaluation long favored in catalyst development laboratories.

We have seen how the combined scanning of flow rate and temperature in a TS-CSTR gives the user unfettered freedom to explore the X-T plane. This is a far cry from the traditional method of catalyst testing using steady state operations. That procedure requires the tedious and time consuming task of bringing the CSTR to some limited number of isothermal steady-state conditions in order to obtain one or a very few points of data. The flexibility and ease of operation of the TS-CSTR promises to make the use of the TS-CSTR much more attractive in catalyst testing, more attractive perhaps than the TS-PFR. Perhaps the inherent advantages of the CSTR configuration will make this type of reactor more common in reaction studies of this kind. It may even encourage the use of CSTRs in mechanistic studies of kinetics.

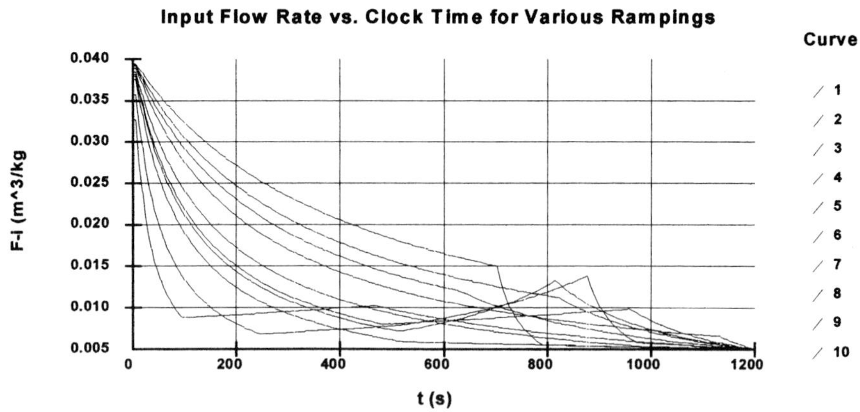

Figure 5.24
Various broken flow ramping rates for the several TS-CSTR runs reported in Figure 5.22. In each case the flow rate was varied continuously from time zero up to the break, at which point the ramping rate was changed.

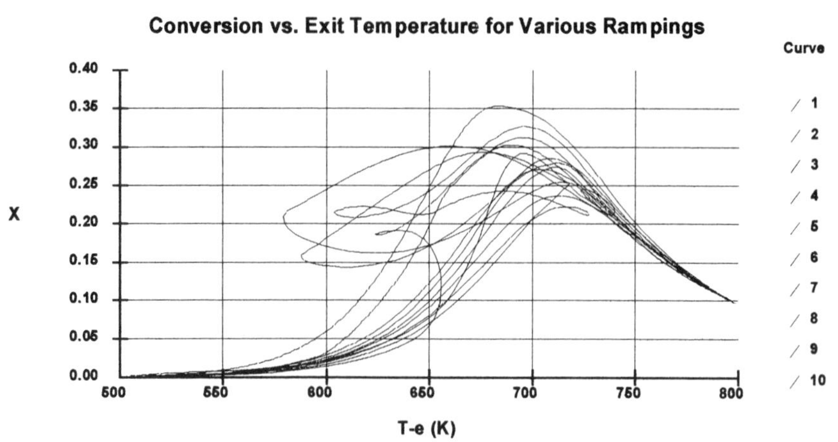

Figure 5.25
The trajectories of the various runs shown in Figure 5.22 remapped to the reaction phase plane. Notice the wide variety of trajectories generated by the relatively random ramping policies. Careful planning can lead to trajectories that visit some particularly interesting region of X-T space.

Flow-Scanning in a TS-PFR

The equations 5.17 to 5.21 governing the TS-PFR all hold instantaneously, and do not at first seem to be affected by any variation in the flow rates the operator may choose to employ. The same overall requirement still holds; however, corresponding plugs in different runs of an experiment all encounter the same conversion and temperature profiles as they transit the reactor. As we will soon see, this will allow the ramping of flow rates in TS-PFR operations, but puts a strict limitation on the methods used in any such ramping.

The easiest way to examine this is to consider once again the hypothetical long multi-port PFR which in a single run encompasses several runs in a fixed length laboratory PFR. As we recall, the outlet conditions of the actual runs, with space times τ_i, are the same as would be measured in the long reactor at positions corresponding to these same space times. If we change the inlet flow rate to this long reactor by some factor we will change the space time values at each of the ports by the same factor. As long as the time scale of clock time is much greater than that of space time, so that the flushing of the reactor takes place essentially instantaneously in comparison to flow rate changes, the conversion readings at these positions will change instantaneously and depend solely on the inlet flow rates. In short, the two times, space time and clock time, are again uncoupled under this condition. In that case, for this single long PFR, the inlet flow rate could be varied freely and rates could be calculated as usual via equation 5.20.

To make this argument more quantitative, consider two conversion readings at space times τ_0 and ½ τ_0. The corresponding positions on the hypothetical long reactor would be some points L and ½ L. Suppose at some time into the run the flow rate into the long reactor were increased by a factor k. Then the space times corresponding to the points L and ½ L would become τ_0/k and ½ τ_0/k. This is equivalent to changing the space times of two real runs, in a fixed length reactor, to those same values, τ_0/k and ½ τ_0/k. In a real PFR this result would be obtained by increasing the flow rate for each of two actual runs by the factor k. In a flow scanning reactor, if the second reading is taken at a space time ½ that at which the first reading was taken, then that same ratio of space times must be preserved throughout the set of data collected at the second flow rate in a TS-PFR experiment. More specifically, if, in the first run of a flow scanning experiment, we chose to increase the flow rate by a factor k at a given clock time, then every subsequent run of the experiment must increase the flow rate by k times at the same clock time.

Generalizing from this example, we see the general condition that must hold with regard to flow scanning in a TS-PFR experiment. As long as the clock time and space time are uncoupled, the first run may have an inlet flow rate $f_1(t)$ that varies in any desired way over clock time t. However, each subsequent run must have a flow rate $f_i(t)$ that is in some fixed proportion to $f_1(t)$: that is, $f_i(t)/f_0(t)$ must be constant as the clock time progresses during each run of the experiment. A more mathematical version of the argument leading to this condition is given in Rice and Wojciechowski, 1997.

Depending on the physical equipment being used, it may or may not be easy to control flow rates precisely along more or less complex trajectories. However, even simple flow schedules, such as linear ramps or step-wise constant ramps, could be advantageous in exploring the X-T plane. Whether the additional complexity involved in the control of a flow-scanning PFR is justified, in view of the advantages offered by a simple temperature scanning arrangement, remains to be seen.

Rate calculations in flow scanning reactors proceed in exactly the same way as outlined above for the temperature scanning case. The only difference is that care must be taken to keep track of the various space times τ_i that are collected with clock time in the data quadruplets (X_i -T_e - t_i - τ_i); they are no longer constant. The resultant sieving is for operating lines, each of which contains points at its own set of space times rather than at the same set of space times for all, as was the case in simple temperature scanning. The effect is as if the sampling ports in the imaginary long reactor were at different positions each time an operating line was constructed. Nevertheless, valid operating lines can be constructed, and from these, valid rate data is calculated.

A Simplified Method of Temperature Scanning

By careful maintenance of isothermality along a PFR, and avoiding excessive space velocities, it is possible to get some of the benefits of temperature scanning in a simplified operation. The procedure involves the ramping of temperature over the desired range while maintaining a constant space velocity. If the ramping rate is such that an increment of feed sees essentially the same temperature as it passes along the reactor, output conversions correspond to those to be expected from a series of isothermal reactors, each operating at a temperature corresponding to that observed at the outlet at the moment of sampling.

In this way, if the rate equation is already known, rates can be calculated at several temperatures from one scan. In effect, each ramping produces a traverse of many isothermal operating lines, as is usual in scanning operations. The difference in this case is that *we know the axial temperature profile for each increment,* and we know it is isothermal. This allows us to treat each measurement as if it were the output of a conventional isothermal reactor, despite the fact that we were in fact slowly ramping the temperature. Even in this simple case the TS procedures described above will allow the data of several rampings to be used to construct the multitude of the actual (they may in fact not be ideally isothermal) operating lines required for the search for an unknown rate equation.

The one-ramp method is strictly applicable only in the case of reactions showing zero volume expansion with temperature and conversion. Inaccuracies creep in when these requirements are not strictly observed. An example of the one-ramp procedure applied to a liquid phase reaction is presented in Chapter 11. In liquid phase reactions, volume expansion normally does not play a role. Moreover, the rate expression of the reaction used in the example given is well established to be first order, making this simple procedure eminently applicable.

The salient point is that this procedure is not general and is prone to error due to imperfections in operation. However, any error that might creep in can be eliminated by doing a full TSR experiment. Then, whether isothermality is maintained or not, whether volume expansion is present or not, treating the results in the way prescribed for TSR will yield correct results. This recommended procedure is sure to give better results than those obtained by straining credulity and treating almost-isothermal data from a single ramp as a series of truly isothermal results.

Interpreting TSR Data Using Integrated Rate Expressions

The possibility of using the integrated form of the rate expression for interpreting TSR data was raised at the end of Chapter 4. The method appears attractive at first sight since it removes the need to take slopes from plots of X vs. t. However, it was pointed out that specific temperature information is required to make this a viable procedure. The information required is the temperature history of each increment of feed during its transformation. This information is obviously available in real time along each TS-BR run (ramping), where the record kept is of conversion-temperature-clock-time (X, T, t) at each data point collected. In the case of the more commonly used TS-PFR, the required information is only available in real-time if the reaction is carried out at near-isothermal conditions. Finally, the TS-CSTR is a differential reactor and there is no need to take slopes from data curves to obtain reaction rates. The integral method offers no advantages in treating data from this type of reactor.

Constraints on the Application of Integral Methods

Even before one considers the need for a temperature profile, restrictions on the applicability of this method start with the integration of the rate expression. Not all rate expressions can be readily integrated or, after integration, give X as an explicit function of the rate parameters. In those relatively rare cases where an explicit integral of the rate equation exists, the integral expression gives X as a function time and of the rate parameters, which in turn are functions of the reaction temperature. For example, in the case of a power-law rate expression there is only one temperature-dependent rate parameter and the integrated rate expression for X is sure to be explicit. However, it is more likely that the integral expression will contain more than one temperature-dependent rate and/or equilibrium constant, and that the dependence between X and the parameters may be in the form of an implicit equation.

In cases where only one temperature-dependent constant appears, it can be calculated in real-time from TS-BR data and, if the reaction is carried out at quasi-isothermal conditions, can also be calculated in real-time from TS-PFR data (see Chapter 11). In all other cases TS-PFR data processing must be done after the experiment is completed.

The necessary information for the integration of the rate expression using TS-PFR data in non-isothermal cases comes from the *operating lines*. Each such line records the X-T-τ information for the increment that entered the reactor at the temperature T_i. By integrating along the operating lines, one can in principle calculate rate constants at temperatures all along the path of each operating line. Unfortunately, if several temperature dependent constants are involved, one must optimize the solution to give *constant Arrhenius parameters* for each of the temperature dependent constants along this path. Then, the same Arrhenius parameters must be applicable along all the other operating lines selected for data fitting. The search for an optimum in these parameters presents a significant computational problem whose complexity may well be discouraging.

The Future of Temperature Scanning

Scanning methods, pre-eminently temperature scanning in kinetic studies, offer a new approach to data gathering in chemical reaction mechanistic investigations. Their advantages over conventional methods promise to be significant. Among these advantages are: temperature control of reactors is much simplified, compared to that for current isothermal requirements; the acquired data is much more plentiful; and, although the basis of temperature scanning is well-rooted in a set of fundamental coupled differential equations, these equations never need to be solved. Instead, data is extracted using simple splining and interpolation and numerical differencing procedures, readily amenable to systematization using discrete methods and digital computers.

To make full use of the potential of temperature scanning methods, a major re-alignment of our current view of kinetic experimentation must be coupled with the availability of modern, fully automated, computer controlled, TS reactors. The process would best begin with universities, where fundamental kinetic studies have traditionally been pursued. Such a development would be sure to revive interest and progress in the study of reaction mechanisms using kinetics.

The large amount of data made available by TS procedures makes it possible, for the first time in kinetic studies, to apply available sophisticated mathematical routines to error correction in raw data, as will be shown in Chapter 7. This, coupled with the productivity of temperature scanning reactors, will make large amounts of *better* data available, and reduce the tedium associated with conventional studies of reaction kinetics. Moreover, the sophistication and variety of issues involved in TSR experimentation is sure to generate renewed challenges and interest, a development that will attract new talent to kinetic studies. How rapidly the promise of scanning techniques will be fully utilized in practice, remains to be seen.

Scanning methods described above need not be confined to temperature or flow scanning in chemical kinetics. There may be a variety of other systems, not at all connected with chemical kinetics, where the concept of variable-scanning can provide a breakthrough method of data acquisition and lead to a resultant advance in understanding.

6. Verification of Kinetic Dominance

Measurements collected in a systematic way will not yield interpretable data unless one is measuring what was intended.

Reaction Rates: Identifying Extraneous Effects

The study of chemical kinetics involves the measurement of the rates of chemical conversion. Unfortunately chemical reaction rates are normally embedded in a sequence of rate phenomena that precede and succeed the actual chemical transformation. In order to be certain that it is the rate of chemical reaction that is being measured, one must ascertain that it is the chemical transformation that is the rate controlling step in the sequence of events associated with that reaction. Some of these intervening rates depend on reactant concentrations, and all rates are temperature dependent.

In a temperature scanning experiment, where both temperature and concentrations vary in each run, one must be sure that the range of conditions that will be covered does not include a transition from a chemical-reaction-controlled regime to one controlled by an intervening process, say diffusion. Validation of temperature scanning data therefore requires careful examination of the conditions present at the site of reaction.

Temperature

We begin with the reaction temperature: at what temperatures is the reaction in the chemical-rate controlled regime? In conventional experimentation, an even temperature in the reaction volume is usually assured by vigorous mixing and waiting for thermal steady state to be established. Vigorous mixing is also a requirement for the same reason in the non-steady state TS-CSTR. However, thermal steady state is normally not established during temperature ramping. Temperature is measured at the reactor outlet, and it is assumed this corresponds to the temperature at which the reaction is taking place in the last increment of a PFR or in the reactive volume of the CSTR. This assumption is usually sound in the case of homogeneous reactions but needs to be examined in catalytic and other fluid/solid processes.

In the TS-PFR it is assumed that, during ramping, the gas phase temperature measured at the outlet of the reactor is the same as that where reaction is taking place, on the surface of the solid involved. This requires careful placement of the temperature sensor just below the last increment of the solid in the PFR bed. The TS method may even benefit from new temperature measuring methods based on optical observation of the solid surface. In two-phase CSTRs, the temperature of the fluid phase most closely approaches the surface temperature of the solid when stirring is vigorous, putting an extra premium on the efficiency of fluid/solid contacting in TS-CSTR design. In both cases, ramping rate will influence temperature equilibration, leading to a tradeoff between the benefits of high ramping rates and the speed, accuracy and cost of available methods of measurement.

Concentration

Similar considerations enter in the case of concentration effects. It is assumed that the reaction is taking place at the concentrations measured at the outlet. This is again always true for homogeneous processes. In fluid/solid reactions there exists the possibility that diffusion gradients exist between the bulk phase concentration and the site of the reaction.

Activity

A further complication arises in the case of heterogeneous catalysis where the activity of the catalyst can change with time on stream or with reactant composition. Changes of this kind are themselves time-dependent processes that are overlaid on the time-dependent kinetics of the chemical transformations. The untangling of all these effects so that pure chemical kinetics can be studied is an important aspect of mechanistic studies using kinetics. (*See* section on catalyst instabilities later in this chapter.)

Testing for Non-Chemical Influences

Attempts have been made to devise calculational methods of testing for the influence of diffusion on the observed rates of reaction. These non-experimental methods for identifying non-chemical influences on reaction rates are highly constrained in their applicability, for example to first order reactions, and often require estimates of hard-to-estimate quantities such as diffusivities. At best these procedures salve one's conscience; more often, they are prone to mislead. The reasons for this are that the mathematics required to estimate the influence of intervening rates are very difficult and only the simplest of cases lead to predictive analytical formulas. In general the only reliable examination of the influence of extraneous effects on the observed rate of reaction consists of an appropriate experimental test.

Experimental Tests for Diffusion Limitations in Conventional Reactors

Homogeneous reactions in fluids rarely involve diffusion limitations. Adequate stirring is normally possible, allowing diffusional effects to be readily eliminated. Diffusional effects can intrude on homogeneous kinetics if, for example, the viscosity of the reactants changes drastically as a result of conversion, as it can in polymerization reactions. In those cases where adequate stirring is not physically possible, or not applied for other reasons, the kinetics of diffusion can dominate reaction rates and these cases represent a separate field of study. In polymerization kinetics, specialized kinetic treatments have been evolved to deal with the circumstances encountered in these reactions. It is not clear, in view of these complications, to what extent temperature scanning will be useful in polymerization studies. The diffusional problems that are most likely to occur in temperature scanning are those to be found in fluid/solid interactions, such as in heterogeneous catalysis.

There are two types of diffusion limitations in fluid/solid interactions: boundary layer diffusion and pore diffusion. Both processes are weakly temperature dependent but the operating conditions that reveal their presence are different.

Boundary layer diffusion

Boundary layer diffusion resistance can be reduced by increasing the linear velocity of the flow passing over the surface. In practice this means increasing the turbulence of the fluid in the presence of the solid particle. To see the effect of turbulence (or fluid linear velocity) at a given space time and the other reaction conditions (temperature, pressure and composition), one has to change the relative velocity between the solid and the fluid.

In the fixed bed fluid/solid CSTR and BR, the recirculation rate must be increased until the measured rate of reaction stops being dependent on this factor. However, if the solid is dispersed in the fluid of the CSTR or BR, consideration must also be given to the problem of slip, which governs the relative velocity between the suspended solid and the suspending fluid. Does additional stirring increase the relative velocity between solid and fluid or does it simply move both of them more vigorously, with minimal effect on the relative velocity?

An example of the configurational change that can be used to examine boundary layer diffusion in the TS-PFR involves a series of reactors of the same volume but varying in their length/diameter ratio. To carry out the diffusion test correctly in a conventional study we first select the reaction conditions, within the range being investigated, where boundary layer diffusion is most likely to be a problem, usually at the highest temperature and longest space time to be used. We then operate the reactor at this condition at steady state and observe the conversion. Next, we replace the reactor with one containing the same amount of catalyst in a vessel with the same active volume but a higher L/D ratio, and observe the conversion. A higher conversion in the second case indicates the presence of bulk diffusion (and perhaps other) constraints in the initial reactor and requires a second iteration with an even higher L/D ratio, and so on, until a constant conversion is obtained. The test conversion should be ~ 50% in the original experiment of each successive pair in order to reduce the effects of experimental error. Duplicate runs should be done to increase confidence in the result. In the end one arrives at an L/D ratio that will assure the absence of bulk diffusion resistance at all the other (less demanding) conditions of the experimental program.

This procedure is fraught with experimental difficulties and the potential for misinformation. It can lead to impossible demands in the reactor configuration or require fundamental redesign of the equipment. It may simply be impossible to carry out the required investigation in practice. For example, an impossible problem arises if the pressure drop through the reactor with the required L/D becomes too great, or if the diameter of the reactor approaches the particle size of the solid. In either case the problem of bulk diffusion will not go away without a remedy. We have simply discovered that kinetic measurements made in that reactor configuration will not yield rates suitable for mechanistic studies. This could be unwelcome news but will save futile effort and the acquisition of misleading information. Fortunately, in most cases, assuring that turbulent flow is maintained, as it must be in a PFR, eliminates boundary layer diffusion barriers, making this type of diffusion the less worrisome of the two.

Pore diffusion

The influence of pore diffusion is more likely to appear and is more troublesome. Its detection therefore requires more vigilance. The experimental test should again be applied at ~50% conversion under the conditions most likely to exhibit diffusion limitations. The procedure involves measuring conversion in a series of identical runs using the same catalyst at successively smaller particle size. When size no longer makes any difference in conversion, the catalyst is free of pore diffusion.

Problems can arise if the particle size required is such that pressure drop through the bed becomes too great. In that case a carefully designed set of confined-fluidized-bed reactors may be used as a source of valid kinetic data, but the solution is less than perfect. And if the fluidized bed is required to remove pore diffusion limitations at the planned conditions, the question arises: is it possible to carry out wide-ranging kinetic studies in reactors of the required configuration? We have already discussed the drawbacks of using confined fluidized beds for kinetic studies in Chapter 1.

Both types of diffusion studies require numerous difficult experiments in conventional reactors, and as a result they are rarely carried out. However, lacking the assurance that the kinetic data is free of extraneous effects, we cannot be sure that it can be reproduced in any other reactor.

Experimental Tests for Diffusion in Temperature Scanning Reactors

In the TS-PFR, diffusion tests rely on the same principles as above. The only difference is that the test involves successive TS-PFR runs over a range of temperatures, at the lowest space velocity planned for the full experiment.

Testing for diffusion using configurational changes

Pore diffusion problems are evidenced by the inability to overlay two runs with different particle sizes in the reaction phase plane. If no diffusional problems exist, the two curves should overlay throughout the temperature range of the ramp (run). If pore diffusion problems appear only at the higher temperatures, one will see the lines overlay at the lower temperatures of the ramp while the larger-particle-size run will begin falling below the small-particle curve as temperature is raised. The TS-PFR procedure has the advantage that it clearly shows the increasing deviation with temperature and can be used to quantify diffusional kinetics as a function of particle size. It can also identify the range of temperatures where diffusion effects are acceptable. The down side is that it requires that a catalyst of different particle size be packed in the reactor after the first run, just as in conventional tests, but the benefit is that just two runs will reveal diffusion effects over the range of temperatures of interest and let us select a safe range of conditions for kinetic studies.

Similarly, a second run using a reactor of different L/D ratio should produce data that overlays the original data in the reaction phase plane if no boundary layer diffusion effects are present. As in the conventional PFR, a change of reactor is required from run to run. However, since the preferred configuration of a TS-PFR involves an oven containing a removable reactor, it is easy to place reactors of various configurations in the oven. (*See* Chapter 13.)

Testing for diffusion using ramping rate changes

An alternative experimental test for diffusion limitations, applicable to TSR operations only, is more efficient in revealing if the particle size intended for use is diffusion limited. It can be carried out without the trouble of changing reactors or crushing the catalyst in search of a diffusion free particle size. This test will show if boundary layer or pore diffusion is present or if the catalyst is affected by aging/activation with time on stream. This TS-PFR-specific test takes advantage of the fact that each of these phenomena can be observed as a function of the temperature ramping rate.

Take the case of pore diffusion. This is a rate process which is time dependent. If we ramp the temperature quickly, then at each position in the bed and at each clock time (and therefore each inlet temperature), the interior of the catalyst may or may not be at steady state with the exterior. If it is not, then a more rapid ramping rate will throw this discrepancy further off steady state, so that the output conversion will not be the same in the two runs. The same will be true for the case of the boundary layer diffusion and for catalyst aging/activation. Fundamentally, therefore, the requirement of temperature scanning is that local steady state in diffusion and adsorption must be achieved rapidly in comparison to the rate at which the inlet temperature is ramped.

Testing for Other Non-Kinetic Influences

To test for all other non-kinetic distortions in the rate measurements the procedure is similar to that suggested above. The first run is performed at the conditions most likely to reveal the problem using an extreme temperature ramping rate. The output conversions are recorded over the range of temperatures scanned. The second run is done under identical reaction conditions but at the other extreme of the intended ramping rate range. If no non-chemical rates interfere with the observed rate of reaction, these two sets of data, when plotted on the reaction phase plane, should be independent of the rate of temperature ramping and therefore should overlay throughout the experiment. Conveniently, the same test will reveal any inadequacies in the rates of measurement defining the range of operation for that reactor. Whether it is reactor inadequacies or diffusion problems that are being observed remains a puzzle for the operator to solve, as always.

The overall conclusion is that TSR ramping rates must be compatible with the rates of all extraneous processes, so that local steady state is established under all conditions of the experiment. In practice the region of ramping rates of interest in TSRs lies between ten and one-tenth of a degree Celsius per minute. More rapid ramping is difficult to achieve in electrically-heated circulating-air ovens, and lower rates make the ramping too slow to make the TSR as efficient as one might wish. The bottom line is that, as long as diffusional or any other extraneous local steady state is established within the time of a few seconds or less, the reaction is amenable to temperature scanning methods of investigation.

By coming to this understanding we have identified a requirement and a restriction of TS-PFR operation. The TS-PFR must be in local steady state with respect to adsorption/desorption, pore diffusion, boundary layer diffusion, etc. There is no equivalent requirement for local steady state with respect to temperature as long as the gas phase temperature measured at the outlet can be identified with the solid surface temperature at which the reaction is taking place.

The situation in the TS-CSTR is similar, with the added concern that recirculation through the bed of solids and gas phase mixing must be adequate at all temperature ramping rates to maintain an instantaneous steady state with the catalyst surface. A similar requirement governs the TS-BR.

Catalyst Instabilities

A number of catalysts change in activity as a result of exposure to reaction conditions. Changes can increase or decrease the activity of the catalyst. They can also be reversible or irreversible. Among the many possible behaviours that this generates, many can be studied using the TS-PFR by applying specific experimental protocols. The four general classes of catalysts whose instabilities can be studied using the TS-PFR are:

1. Catalysts whose activity increases fairly rapidly until they achieve a constant final state that remains stable over the range of reaction conditions used for the duration of a TS-PFR experiment.

Such catalysts should be stabilized before kinetic experimentation begins. Subsequent experiments reveal the kinetics of the stabilized catalyst. Different methods of stabilization reveal kinetic differences between samples and the mechanistic implications of the mode of pretreatment.

2. Catalysts that increase fairly slowly in activity but finally reach a stable steady condition as in case 1 above.

Catalysts of this type may simply be stabilized until they reach a constant activity or they may merit having their activation processes studied. "Snapshots" of the activity at various stages of activation can be taken by doing TS-PFR experiments at various stages of activation to reveal changes in the rate constants that accompany the activity increases.

Those catalysts whose kinetics of *activation* are to be studied have to display time-on-stream-dependent kinetics of activation if these are to be studied in a TS-PFR. At this time no other kinetics of activation can be unraveled using available TS-PFR theory. The procedures to be used are similar to those described below in the section on catalyst decay.

3. Catalysts whose activity decays so slowly that no perceptible change takes place during a TS-PFR experiment (i.e. a set of ramping runs).

Slowly decaying catalysts can be examined at a sequence of activities by performing an experiment at successive stages of decay. The decay itself takes place in some auxiliary aging unit. The changes in the kinetic parameters of a mechanistic rate expression will then reveal what exactly is changing in the mechanism of the reaction. This procedure is identical to that suggested for slowly activating catalysts.

4. Rapidly decaying catalysts, ones that show a partial, not total, loss of activity during each *run.*

The kinetics of decay of such catalysts, together with the kinetics of the completely fresh catalyst, can be studied using a TS-PFR as long as the decay is simply a function of time on stream.

To sum up: catalysts exhibiting rapid activity changes (up or down) can be studied using the TS-PFR as long as the changes of activity are purely a function of time-on-stream and temperature. Suitable procedures are discussed below. Influences of other factors, such as reactant composition or poisoning species concentration, have to be absent or held constant in each experiment. A more detailed discussion may be found in Grenier (1997).

Quantifying Catalyst Decay

Catalysts are often, if not always, subject to changes in activity during their time on stream. These changes range from increases in activity during early times on stream, while the catalyst is being activated, to rapid loss of activity in the presence of the feed, as in the case of cracking catalysts. Most of these changes depend on the environment surrounding the catalyst. In cases where the level of conversion is a factor, the quantification of these changes can be difficult, but if the environment is reasonably constant and the activity changes are simply a function of time on stream, the situation is much simpler.

Fortunately, as long as the environment is kept reasonably constant, many catalysts do change activity simply as a function of time on stream. The influence of temperature changes in such cases is also readily accommodated. The seminal idea behind the treatment of such time-dependent activity changes is to consider the change of activity as a kinetic process where the catalyst activity is changing as if the catalyst were in a batch reactor, in the presence of constant reaction conditions, regardless of the actual configuration of the reactor where the catalyst resides.

A more detailed interpretation of this idea is that the number of active sites is changing as a function of time on stream, which is a measure of the time the catalyst spends in the above mentioned batch reactor. This behaviour is quantified by using a power law to describe the kinetics of activity change. Power-law kinetics have in fact been shown (see Kemp and Wojciechowski, 1974) to apply in all cases where multiple parallel first order processes lead to changes, in this case to decay or activation. As decay is often the result of multiple simultaneous causes, the kinetics of catalyst decay in catalytic cracking, and other cases, follow a simple power-law behaviour despite the fact that the actual decay mechanism is quite complicated. The use of a power-law rate expression to quantify activity changes is therefore not a simple heuristic approximation but is justified by a rigorous mathematical derivation of the kinetics involved and by experimental investigations of sets of parallel reactions. To a good approximation, the overall rates of large sets of parallel reactions, whether in decay or elsewhere, obey power-law kinetics.

The rate of decay is therefore written:

$$-r_d = k_d\theta^m = -\frac{d\theta}{dt} \tag{6.1}$$

where t is time on stream (clock time),
θ is the fraction of catalytic sites active at time t,
k_d is the decay rate constant,
m is the order of the site loss reaction with respect to site concentration.

The integrated form of this expression gives the fraction of sites active at any time on stream, t:

$$\theta = \frac{1}{[1+(m-1)k_d t)]^{1/(m-1)}} \tag{6.2}$$

It is generally true that a catalytic rate expression is of the form:

$$-r_A = f(C_A, \theta^n) = \frac{dC_A}{dt} \tag{6.3}$$

where $f(C_A)$ is a function of the concentration of the reactants. Its specific form depends on the mechanism of the reaction;
n is the order of the catalytic process with respect to the concentration of active sites;
τ is the space time for the reactants.

The rate of a catalytic reaction therefore depends on two times: the time of contact between the reactants and the catalyst (τ) and the time that the catalyst has been in the presence of activity-altering conditions (t). As long as these two times are very different they are "uncoupled" and one can integrate the conversion expression, assuming constant t, to obtain the instantaneous concentration of unconverted feed, $C_{A\ inst}$. If necessary one can then integrate and average the result with respect to the time on stream, t. This yields the average conversion $C_{A\ avg}$ over a period of time while the catalyst was decaying.

$$C_{A\ inst}(\tau, t) = f(C_A(\tau))g(\theta(t)) \tag{6.4}$$

$$C_{A\ avg}(\tau, t) = \frac{f(C_A(\tau))}{t_f - t_i}\int_{t_i}^{t_f} g(\theta(t))dt \tag{6.5}$$

Identical procedures will deal with activity increases during catalyst activation. All activity changes are readily treated as kinetic phenomena as long as they are simple functions of time on stream. In such cases the TS-PFR can be used to quantify both the kinetics of the catalytic reaction and those of the accompanying activity changes.

Studies of catalyst activity using the TS-PFR

In the TS-PFR an increment of feed, as it proceeds along the reactor axis, passes over the catalyst at various temperatures. Regardless of this, as previously discussed, the rate of reaction is measured only at the outlet of the reactor, and at the outlet temperature. The same procedure will yield rates on a decaying catalyst. The measured rates will, however, be different from those that would have been observed if the catalyst was not decaying. To understand this we have to re-examine the defining equations of the TS-PFR.

We begin by noting that activity at any time t can be defined as:

$$a(t) = \frac{r(t)}{r(0)} = \theta^n \qquad (6.6)$$

where $a(t)$ is the activity of the catalyst at time t;
$r(t)$ is the rate of reaction at time t;
$r(0)$ is the rate of the reaction that would have been observed at time t if no decay had taken place;
θ is the fraction of the original sites still active at time t;
n is the number of sites required by the rate-controlling step in the reaction. In most cases $n = 1$.

Since activity is directly related to the number of sites available for reaction we identify current activity at time t, as defined above, with the fraction of sites still active:

$$a(t) = \theta^n(t) = \frac{1}{(1+(m-1)k_d t)^{\frac{n}{m-1}}} = \frac{1}{(1+Gt)^N} \qquad (6.7)$$

where k_d has the usual Arrhenius form:

k_d = A exp(−B/T)
G = $(m-1)k_d$;
N = $n/(m-1)$;
θ = the fraction of original sites still active at time t.

When we include this concept in the set of equations describing the TS-PFR (see Chapter 5) we obtain the following;

$$\frac{\partial X(s,t)}{\partial s} = -\frac{r(X,T)\delta(X,T)}{C_{A0}} \qquad (6.8)$$

$$\frac{\partial T(s,t)}{\partial s} = -k_1\delta r(X,T) - k_2\delta[T(s,t) - T_R(s,t)] - k_3\delta[T(s,t) - T_C(s,t)] \qquad (6.9)$$

$$\frac{\partial T_R(s,t)}{\partial t} = -k_4[T(s,t) - T_R(s,t)] - k_5[T_R(s,t) - T_E(t+s)] \tag{6.10}$$

$$\frac{\partial T_C(s,t)}{\partial t} = -k_6[T(s,t) - T_C(s,t)] \tag{6.11}$$

This set of equations is solved under the boundary conditions:

$$\begin{aligned} T_R(s,0) &= T_C(s,0) = T_E(0) \\ X(0,t) &= 0 \\ T(0,t) &= T_i(t) \\ a(s,0) &= 1 \end{aligned}$$

This allows the calculation of rates, as modified by activation or decay, from data at the exit of the reactor. By starting each run with a sample of fully active catalyst, whether due to regeneration or replacement, the above relationships allow us to calculate sets of data corresponding to r(t) just as we were able to do in the case of the r(0) rates discussed previously.

Rates of this kind are of relatively little use, as it is impossible to sieve out isothermal sets of rates at constant activity. It would take many experiments at different conditions to make available a sufficient data set to allow the sieving out of isothermal data at constant activity. Such data would, however, allow us to compare sets of isothermal rates at a variety of activities and from this determine the parameters of the deactivation equation. Unfortunately, the number of experiments required for this procedure to be applied presents a serious handicap. The method cannot be made compatible with rapid data acquisition. Fortunately, there is a way out of the dilemma. It is possible to unravel decay properties by other much simpler and less laborious means. The method described below (see also Grenier, (1997)) requires as few as two experiments using the decaying catalyst.

The two experiments are carried out at different ramping rates. In fact either experiment can be carried out with a broken ramp consisting of a fixed period of ramping at one rate followed by a second period at a higher or lower ramping rate. The ramping must be such that the data, when plotted on the reaction phase plane of X vs. T_e, will result in traces of individual runs that cross one another when the two experiments are overlaid.

Figure 6.1 shows an example of two such experiments. The use of a broken ramp in one of the experiments leads to crossovers that are more orthogonal, so that the crossing points are easier to identify precisely. As we will see, this makes subsequent calculations more accurate. Notice also that each curve of experiment 2 can cross several curves of experiment 1. This is important as it is the cross-over points that supply the required information about catalyst decay. The more cross-over points one can generate between the two experiments the more data is made available for further manipulation and interpretation.

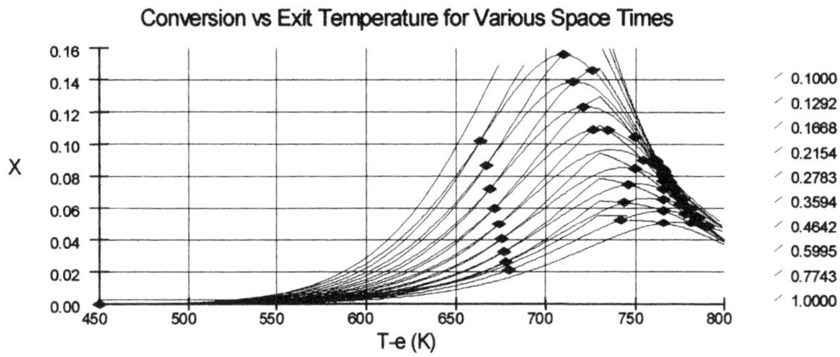

Figure 6.1
An overlay of two experiments designed to evaluate catalyst changes occurring as a function of time on stream. The first experiment consisted of rampings similar to those shown in Figure 5.4. The second experiment consists of a series of runs with broken temperature ramps designed so that the trajectories of the two experiments cross over. The crossing points are identified by diamonds.

This simple procedure for studying catalyst decay is based on the understanding that the rate calculated in a TS-PFR experiment refers to processes taking place in the increment of catalyst at the very end of the reactor. At that point we know the temperature history of the catalyst at each instant of clock time during each run. As clock time progresses the catalyst ages as a function of time but the decay constant also changes due to changing temperatures. Since decay is a cumulative process the extent of decay at the end of the bed will depend on the thermal history of the end-increment. To calculate this history we need again to integrate equation 6.1, but now treating k_d as a variable over time. This can be achieved by a numerical integration of k_d with the observed behaviour of T_e, the exit temperature of the reactor.

We obtain the value of G_{avg} by a numerical integration of $k_d(T_e)$ using the observed time-behaviour of T_e, the exit temperature of the reactor. Since the exit temperature T_e is a function of clock time t, the integration can be done in the time domain.

$$G_{avg} = \left(\frac{m-1}{t}\right)\int_0^t k_d(t)dt \tag{6.12}$$

We can now return to the definition of activity and substitute the average time-on-stream activity function to obtain

$$r(t)_{(X,T)} = \frac{r(0)_{(X,T)}}{(1+G_{avg}t)^N} \tag{6.13}$$

From the two experiments described above, a number of cross-over points is available. These points contain information concerning the rate and the clock time at that point for each curve separately. Rewriting the activity relationship in equation 6.13 in logarithmic terms gives:

$$\ln(r(t)_{(X,T)}) = \ln(r(0)_{(X,T)}) - N\ln(1+G_{avg}t) \quad (6.14)$$

whence we see that the rate at such a point, in the form ln(r(t)), is a linear function of clock time in the factor $\ln(1+G_{avg}t)$, with a slope of –N. This is true for rates at each point on the reaction phase plane. More importantly, it is true at each intersection point, as well as at each point along the intersecting curves.

If we postulate initial estimates of values for the exponential term B and for the term $A' = (m-1)A$ in the parameter $G = A'\exp(-B/T)$, we can perform the numerical integration described above and obtain the pertinent values of G_{avg} from equation 6.12. Using these we plot the logarithm of each of the rates at the intersection point against $\ln(1+G_{avg}t)$.

The difference between the two rate values at a cross-over from the two experiments is solely due to decay, since both rates are measured at the same conversion (X) and temperature (T_e). According to the above logarithmic relationship the two points therefore belong on a straight line with slope N and rate-axis intercept of $r(0)_{(X,T)}$. The same is true for each pair of rates at all the cross-over points. This in turn means that all such pairs should lie on parallel lines in the $\ln(r(t)_{(X,T)})$ vs. $\ln(1+G_{avg}t)$ plane, since the slope defined by the decay exponent, N, is the same for all pairs.

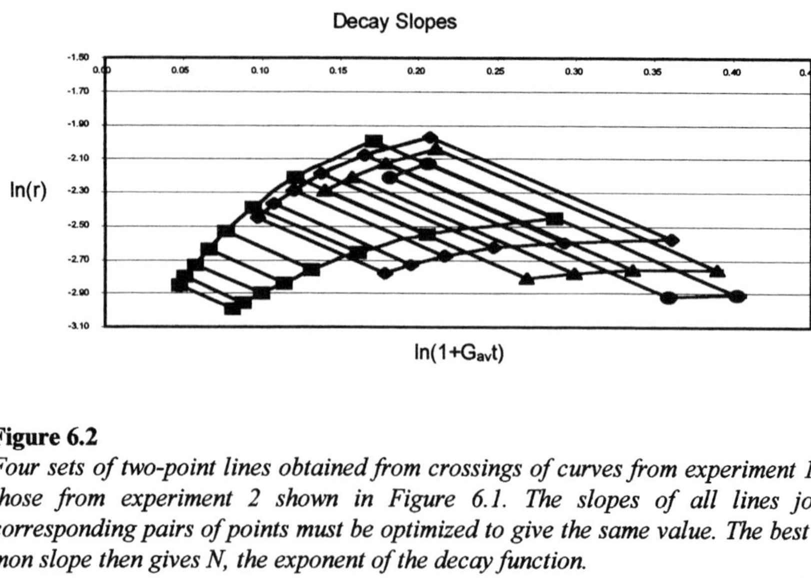

Figure 6.2

Four sets of two-point lines obtained from crossings of curves from experiment 1 with those from experiment 2 shown in Figure 6.1. The slopes of all lines joining corresponding pairs of points must be optimized to give the same value. The best common slope then gives N, the exponent of the decay function.

Figure 6.2 shows the expected behaviour drawn from a hypothetical example. The expected parallel lines are seen to join pairs of points from each intersection. In this example there are four sets of such pairs, identified by the shape of the data points in Figure 6.2, each set belonging to a set of curves from experiment 1 in Figure 6.1 crossing curves from experiment 2 in the same figure. There are a number of such crossings possible since curves at a given τ from experiment 1 can cross curves at τ, 2τ and so on, in experiment 2. Four sets of such crossings are shown in Figure 6.2

In this way we have developed a criterion for solving the decay problem. We optimize our estimates of B and A′ to minimize the dispersion, δ, of the values of the slopes of the lines on the $\ln(r(t)_{(X,T)})$ vs. $\ln(1+G_{avg}t)$ plane:

$$\text{minimize } \delta = \sum_i \frac{(N_i - N_{avg})^2}{n} \tag{6.15}$$

Once N is well determined, we automatically have the value for B = E/R, where E is the activation energy for the decay. The decay frequency factor is A = A′N/n with n normally taken to be 1.

This allows us to write a quantitative expression for the rate of catalyst decay in this system. With this information we can purge both of the experiments we already have, removing decay influences so that we obtain all the rate data already established in the two experiments in a form suitable for fitting a decay-free rate expression for the reaction. Having two experiments presents an opportunity for extensive crosscheck of our rate expressions. We also note that we have a set of zero decay rates from the intercepts $r(0)_{(X,T)}$. This set is usually too small to be used for rate parameter estimation by itself but can be used as a further cross check, at conditions of special interest to theoreticians, on the parameters of the rate expression for the purged rates from the two experiments.

Using three or more experiments

Although the data for decay rate evaluation from two experiments is sparse in terms of the volumes of data available from normal TSR experiments, it can be significantly increased by using more experiments than the required minimum of two. The reason is that the number of available crossing points quickly increases with the number of experiments performed. As a rough estimate of the rate of increase of available data we write:

$$M = \frac{n!}{2(n-2)!}$$

where M is the multiplier to be applied to the number of points that were made available from the first two experiments, and n is the number of experiments done. For example, by doing six experiments, corresponding to three times the effort required to obtain the minimum of two experiments, we obtain fifteen times as much data suitable for evaluating the decay behaviour of the catalyst as we would have obtained from the required two experiments.

Interpolation

Figures 6.1 and 6.2 show that the number of cross-over points directly available from two experiments is far sparser than the number of points available in a standard TSR experiment and therefore the subsequent calculations will lead to greater uncertainty in parameter estimates. It is possible, however, to employ an interpolation scheme to manufacture new virtual crossing points. This would seem to allow an unlimited increase in the number of points available, but it does not really provide any new information, and will not significantly help parameter estimation in the two-experiment case. However, such an interpolation scheme does allow for two advantageous possibilities.

1. Since the exit (X, T) data for each run are recorded at discrete times rather than continuously, it is likely that a crossing point will occur between recorded data points. Usually the recorded points are close enough together that this slight error will not have a great effect. Nevertheless, enhanced precision can be obtained by interpolating the data to the exact crossing point.

2. If data from three experiments is available, then we might look for triple crossing points where curves from all three experiments cross simultaneously. It is very unlikely that such triple crossings will occur spontaneously, but we can interpolate data from the third experiment to pass through a crossing point from the first two experiments. This would generate a "virtual triple crossing point". This procedure might well lead to significant improvement in parameter estimates.

We describe below a simple but satisfactory interpolation scheme, and then discuss in more detail how it might be used.

The triple point interpolation

The hallmark of temperature scanning is the availability of sophisticated methods of data processing. These are predicated on the availability of large and consistent datasets that result from TSR operations. In the case of activity studies, various methods can be used to augment the original data by mathematical manipulation. One such method is to seek interpolations that yield triple point intersections for use in calculating the decay constants.

Let P_0 be an arbitrary point in the X-T plane, and suppose it lies between curves C_1 and C_2 from, say, the first experiment (see Figure 6.3). We wish to interpolate a curve between C_1 and C_2 passing through P_0. Each of the curves C_1, C_2 can be thought of as a parametric curve in terms of clock time t: i.e. $C_i(t) = (T_i(t), X_i(t))$, where $T_i(t)$ and $X_i(t)$ are the exit temperature and conversion at clock time t. Let P_1, P_2 be points on C_1, C_2 such that the line segment joining them passes through P_0 and is as short as possible. Since the curves C_1, C_2 are nearly parallel, this line segment will be nearly perpendicular to C_1, C_2. Let t_1, t_2 be the clock times at which these points are reached on their respective curves, and let λ be the interpolation parameter value for which

$$P_0 = (1-\lambda)P_1(t_1) + \lambda P_2(t_2) \tag{6.16}$$

Then the clock time corresponding to P_0 is (at least approximately)

$$t_0 = (1-\lambda)t_1 + \lambda t_2 \tag{6.17}$$

For any time $0 \leq t \leq t_0$ we may define $k = t/t_0$, the fraction of the way along the curve, and then define the interpolated curve as

$$C(t) = (T(t), X(t)) = (1-\lambda)C_1(kt_1) + \lambda C_2(kt_2) \tag{6.18}$$

This interpolates a temperature history for the point P_0, and allows calculation of the necessary integral to obtain $\ln(1+G_{avg}t_0)$.

We observe that it is not strictly necessary to compute both components of the interpolated curve. Once λ and t_0 have been found it is slightly (but only slightly) less work to compute only the necessary interpolated temperature history:

$$T(t) = (1-\lambda)T_1(kt_1) + \lambda T_2(kt_2) \tag{6.19}$$

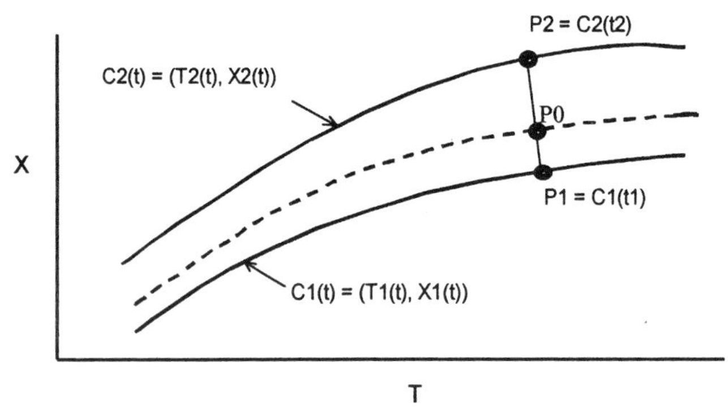

Figure 6.3
C1 and C2 are two X-T curves from an experiment. P0 is an arbitrary point between them. The dotted curve is an interpolated curve for the experiment.

This simple linear interpolation is quite coarse, and certainly far too crude to be useful for calculating rates, which involves estimating slopes and hence is very sensitive to small errors. For rate estimation it is necessary to use a much more elaborate interpolation scheme that is described in Chapter 7. For our present purposes, however, we are only using the interpolated curve to estimate the integral needed to calculate G_{avg}, and this is quite insensitive to small errors. Consequently the simple interpolation scheme outlined above is satisfactory for this purpose.

We may make use of such interpolations in a variety of ways. First, as noted previously, it is likely that a crossing point will occur between recorded data points. Usually the recorded points are close enough together that this slight error will not have a great effect. Nevertheless, enhanced precision can be obtained by interpolating the data to the exact crossing point.

Second, observe from Figure 6.3 that the crossing points lie on families of curves, call them X-curves, which will run smoothly through these points if the X-T curves from the various runs are evenly spaced, as is usually the case. Irregularities in these curves then correspond to errors in the observed locations of the crossing points. This can easily arise, for instance, for points where two X-T curves cross at a very small angle. It may then be advantageous to smooth the X-curves, and use the smoothed curves to locate crossing points through which X-T can then be interpolated.

Finally, when data from three experiments is available, we may construct triple crossing points, where curves from all three experiments cross simultaneously. To calculate such triple crossing points we need merely choose a crossing point P_0 for the first two experiments, identify curves C_1 and C_2 from the third experiment that have P_0 between them (as seen in Figure 6.3), and interpolate a curve between C_1 and C_2 passing through P_0. Applying the same logarithmic relationship as above to such a triple crossing point results in three points in the $\ln(r(t)_{(X,T)})$ vs. $\ln(1+G_{avg}t)$ plane that should all lie along a straight line with slope $-N$. The criterion for estimating this slope would still be to minimize the dispersion of the slopes of all the line segments, but extra weight could be put on equalizing the slopes of line segments in the triple crossing lines.

If data from more than three experiments is available, the same interpolation scheme can be used to produce higher-order crossing points. The work of interpolation needed to produce these multiple crossing points may or may not be justified in terms of improvement in the parameter estimates. It may be that the simpler scheme suggested above of just using crossing points for pairs of X-T curves will give sufficiently good results. A better understanding of these considerations is required, but it would seem that, if appropriate algorithms are developed to carry out the required calculations, computers will make short work of this otherwise daunting problem.

It may also be true that taking crossing points from just two runs at a time and treating them in the prescribed way will prove to be the most robust method of optimizing the decay parameters.

A Suggestion

Every attempt at identifying the mechanism of a chemical reaction should be preceded by a verification of the fact that rate measurements are being made under conditions where the chemical reaction is the rate governing step. Studies that report no such verification, or present no good argument for chemical reaction being the rate limiting step, are of little value as archival sources of information and do not deserve to be cited. Moreover, conclusions drawn on the basis of such studies are also dubious. The volume of literature in the field of chemical kinetics, as well as the number of citations, would be usefully reduced if a policy requiring such proof was instituted by the editors and reviewers of the kinetics literature.

7. Processing of Data

The data that is collected is a-buzz with noise and subject to systematic distortion. To understand the message hidden within, one must separate it from the surrounding noise and confusion.

Introduction

This chapter and the next discuss the steps involved in passing from raw experimental TSR data to a final fitted rate expression. This chapter deals with the processing necessary to extract conversion and rate values from the noisy raw data. Chapter 8 describes various methods for fitting such conversion-rate data to rate expressions.

A TSR experiment will normally produce mole fraction measurements of the various components in the outlet stream. Inlet stream components could also be measured at each instant, but are generally presumed known and constant, once the calibration and set-up for a given experiment has been done. The outlet measurements will be subject to random error due to uncontrollable instrument noise, measurement error, etc. There is a temptation to smooth this data at once, in the hopes of eliminating some of the noise, but this is unnecessary at this stage, and actually counterproductive, since a much more delicate smoothing is required later, in three dimensions, when the slopes that represent the rates of reaction are estimated.

However, there is one adjustment of the outlet stream data that is useful at this first stage. Because of noise in the data, the measured mole fractions in the outlet stream will probably not be in atomic balance with the inlet stream. The first section of this chapter describes how to make minor corrections to the outlet molar composition in order to achieve this physically necessary mass balance. As we will see below, there is a large bonus attached to this procedure

Subsequent sections of this chapter discuss procedures for calculating fractional conversions X of the various feeds and products. The main difficulty here is in dealing with volume expansion. In traditional steady-state experiments, volume expansion is usually handled by using a relatively straightforward expansion coefficient ϵ based on the stoichiometry (known or measured) of the reaction. In TSR experiments this is much more difficult since changing conditions during the experiment may change the stoichiometry in unknown ways. To deal with this, procedures are developed for computing a different (variable) expansion coefficient δ that is recommended for TSR experiments.

Finally, a completely different approach to dealing with volume expansion is proposed. This procedure yields a value for δ with each analysis, is much simpler to use, and avoids altogether the need to approximate expansion using the expansion coefficient ϵ. With fractional conversions X computed, the next step is to compute rates $dX/d\tau$. This is a delicate procedure, because errors in measured X values can lead to large errors in the measured slopes. It turns out that simply fitting surfaces to the conversion-temperature-space-time data and then calculating slopes by numerical differences leads to unacceptably poor estimates of the rates. What is required instead is a fitting procedure that simultaneously approximates the data and the slopes. Fitting by *basic splines* (B-splines) turns out to give very satisfactory results. This is described in the final sections of the present chapter.

With conversion-temperature-rate data in hand, Chapter 8 proceeds to discuss fitting rate expressions to this data. The approach is to estimate rate expression parameters using nonlinear least squares fitting. This requires the use of nonlinear optimization techniques, of which there is a considerable variety available. Chapter 8 includes a brief summary describing a number of these optimization algorithms.

Chapter 8 concludes with an overview of a general parameter fitting procedure suitable for TSR experiments. In following the principles described there, it is desirable to keep in mind the various scaling and transformation techniques described in Chapter 4, that can considerably improve the final fitted results.[1]

Transforming Analytical Results for Data Fitting

Adjusting for Atomic Balance

The results obtained from analytical equipment measuring the composition of a reactor output are normally reported in mol fractions. In most cases the mol fractions of components at the outlet are automatically re-normalized so that their sum is one. This is correct as far as mol fractions are concerned, but the atomic balances of the output may well be off due to both analytical error and the process of re-normalization applied to the measured mole fractions. This is an issue that needs to be addressed in each instance, regardless of reactor configuration or the mode of reactor operation. At the same time this is a particularly well conditioned problem to address in the case of the voluminous data from a TSR experiment.

Because of the plentiful data available from a TSR experiment it is possible to rationalize the reported output composition by seeking a result where both the molar composition and the atomic balances are such that their sums present an optimum solution. The procedure to achieve this under a variety of circumstances is as follows.

The definitions

Assume the following notation:

- $\boldsymbol{\alpha}$ is the initial concentration vector (i.e. at zero conversion, in mol fractions in the input stream);
- $\mathbf{c}$ is the initial estimate of the current output composition vector;
- $\mathbf{M}$ is a mass spectrometer calibration matrix;
- $\mathbf{v}$ is a vector of measured mass spectrometer peaks for the total spectrum of the output a reactor;
- $\mathbf{A}$ is the atomic component matrix for the reacting mixture;
- $\boldsymbol{\beta}$ is the output composition vector in moles per mol of total input.

[1] Software designed to carry out the data treatment described below, as well as for controlling and data logging some TS Reactor configurations, can be obtained from the licensed distributor, SE Reactors Inc., 6 Strathearn Blvd., Toronto Ontario, M5P 1S7, Canada; phone/fax 416 483 4234 or 613 267 4942.

Suppose we have a chemical reaction involving n elements $E^1, E^2, \ldots, E^n$, making up m different molecules $M_1, \ldots M_m$ where molecule M_j has a_{ij} atoms of element E^i. Suppose some quantity of feed has α_j moles of M_j, and the product from this quantity of feed has β_j moles of M_j. The chemical reaction (stoichiometric) equation for this process is:

$$\begin{aligned}&\alpha_1 E^1_{a_{11}} E^2_{a_{21}} .. E^n_{a_{n1}} + \alpha_2 E^1_{a_{12}} E^2_{a_{22}} .. E^n_{a_{n2}} + \ldots + \alpha_m E^1_{a_{1m}} E^2_{a_{2m}} .. E^n_{a_{nm}} \\ &= \beta_1 E^1_{a_{11}} E^2_{a_{21}} .. E^n_{a_{n1}} + \beta_2 E^1_{a_{12}} E^2_{a_{22}} .. E^n_{a_{n2}} + \ldots + \beta_m E^1_{a_{1m}} E^2_{a_{2m}} .. E^n_{a_{nm}}\end{aligned} \tag{7.1}$$

For the chemical reaction to be balance in atomic content we must have:

$$\sum_j \alpha_j a_{ij} = \sum_j \beta_j a_{ij} \;, \quad i = 1, \ldots n \tag{7.2}$$

This gives n equations, one for each element, that we may write as a matrix equation:

$$\mathbf{A\alpha} = \mathbf{A\beta} \tag{7.3}$$

for the matrix $\mathbf{A} = [a_{ij}]$ and the vectors $\boldsymbol{\alpha} = \begin{bmatrix} \alpha_1 \\ \alpha_2 \\ . \\ . \\ \alpha_m \end{bmatrix}, \ \boldsymbol{\beta} = \begin{bmatrix} \beta_1 \\ \beta_2 \\ . \\ . \\ \beta_m \end{bmatrix}$

We may suppose we are considering one mole of feed, or that the input vector α is in partial pressures so that its components are in the mole fractions directly. Due to volume expansion caused by reaction, temperature, and pressure variations, the outlet molar composition vector $\boldsymbol{\beta}_i$ will be the same as the outlet partial pressures but scaled by some expansion factor. We will see what that factor is after we consider fractional conversion and yields.

The cases to be considered

In the following sections we first consider the case where the outlet mole fractions or partial pressures are measured directly, for example by a gas chromatograph, but due to noise or measurement error need to be adjusted in order to achieve the atomic balance required by equation 7.3. Next we consider the case of a mass spectrometer, where calibration is required to derive the partial pressures from mass spectrometer peak readings. In that case it turns out that the procedure used to achieve mass balance also provides a useful way to improve the calibration and to monitor possible changes (calibration drifts) in the mass spectrometer. Finally we consider the case where some components in the feed and/or product are not measured, due either to ignorance or to neglect. It turns out that even in this case it is possible to calculate good approximations to the outlet molar composition to be expected in order to maintain mass and atomic balance.

In all the procedures mentioned above and aimed at achieving good approximations while maintaining mass balance, it is necessary to find least-squares solutions to systems of linear equations subject to linear constraints. The mathematics for doing this is described at the end of this chapter.

Case 1. Outlet partial pressures are directly available.

Denote the partial outlet pressures by:

$$\mathbf{p} = \begin{bmatrix} p_1 \\ p_2 \\ . \\ . \\ p_m \end{bmatrix}$$

We suppose these are measured but are subject to error. We wish to find the vector of mole outputs $\boldsymbol{\beta}$ and a constant k so that

$$\boldsymbol{\beta} \cong k\mathbf{p} \tag{7.4}$$

subject to the mass balance equation 7.3. We first estimate k, by observing that from equations 7.3 and 7.4 we have

$$\mathbf{A}\boldsymbol{\alpha} \cong k\mathbf{A}\mathbf{p} \tag{7.5}$$

The least-squares approximate solution to this equation is

$$k = \frac{(\mathbf{Ap})^t(\mathbf{A}\boldsymbol{\alpha})}{(\mathbf{Ap})^t(\mathbf{Ap})} \tag{7.6}$$

For a proof of this and all the other mathematical results in this and the next three sections, see the mathematics section at the end of the chapter.

Using this value of k, we may now find the least-squares approximate solution $\boldsymbol{\beta}$ to equation 7.4 giving the outlet molar contents subject to the mass balance condition in equation 7.3. The solution turns out to be

$$\boldsymbol{\beta} = \boldsymbol{\alpha} + \mathbf{N}(\mathbf{N}^t\mathbf{N})^{-1}\mathbf{N}^t(k\mathbf{p} - \boldsymbol{\alpha}) \tag{7.7}$$

where **N** is a *null matrix* for **A**, i.e. a matrix whose columns form a basis for the null space of **A**, or, in other words, a matrix with linearly independent columns and such that every vector **u** satisfying **Au** = 0 can be written as **u** = **Nz** for some **z**. A method for calculating **N** is given in the mathematics section at the end of this chapter.

Case 2. Outlet partial pressures are derived from mass spectrometer data

Let the vector **v** denote the outlet *mass-spectrometer peak-data*, and let **M** denote the *calibration matrix* for the various elements. That is, each column of **M** gives the calibrated peak-data for one component, and the rows correspond to some given set of atomic mass numbers. **M** need not be square, but it must have at least as many rows (atomic masses) as columns (molecular components). With each analysis there is a *constituent vector* **c**, i.e. some multiple of the output partial pressure vector **p**, such that (assuming perfect calibration):

$$\mathbf{M}\,\mathbf{c} = \mathbf{v} \tag{7.8}$$

If **M** is square, we can calculate $\mathbf{c} = \mathbf{M}^{-1}\mathbf{v}$. If **M** is not square we may use the least-squares solution. In any case, since v and **M** likely have experimental error, equation 7.8 will likely not yield **c** as an exact multiple of the true **p**. Nevertheless, we may estimate **c** as the least-squares solution to equation 7.8:

$$\mathbf{c} = (\mathbf{M}^{t}\mathbf{M})^{-1}\mathbf{M}^{t}\mathbf{v} \tag{7.9}$$

We can now normalize **c** by dividing by $\Sigma \mathbf{c}_i$ to get an approximation to the outlet partial pressures **p**, and then proceed exactly as in the previous section to estimate $\boldsymbol{\beta}$. A better procedure, however, one that involves fewer levels of approximation, is to observe that **c** is approximately a multiple of **p** and **p** is in turn a multiple of $\boldsymbol{\beta}$. From this we have:

$$\boldsymbol{\beta} \cong \mathrm{k}\,\mathbf{c} \tag{7.10}$$

for some k to be determined. Substituting this into equation 7.8 yields

$$\mathbf{M}\boldsymbol{\beta} \cong \mathrm{k}\,\mathbf{v} \tag{7.11}$$

Thus we may estimate $\boldsymbol{\beta}$ as the least-squares solution to equation 7.11, subject, as always, to the mass balance equation 7.3. To do this, we first estimate k just as before, observing that equation 7.10 is exactly analogous to equation 7.4, to get

$$\mathrm{k} = \frac{(\mathbf{Ac})^{t}(\mathbf{A}\boldsymbol{\alpha})}{(\mathbf{Ac})^{t}(\mathbf{Ac})} \tag{7.12}$$

and then calculate the best approximate solution to equation 7.11 subject to the mass balance equation 7.3 as follows.

Let **N** be a null matrix whose columns form a basis for the null space of **A**. That is, $\mathbf{AN} = \mathbf{0}$ and every solution to $\mathbf{Ax} = \mathbf{0}$ can be written uniquely in the form $\mathbf{x} = \mathbf{Nz}$ for some **z**. Let $\mathbf{L} = \mathbf{MN}$, and let $\mathbf{L}^{+} = (\mathbf{L}^{t}\mathbf{L})^{-1}\mathbf{L}^{t}$. Then

$$\boldsymbol{\beta} = \boldsymbol{\alpha} + \mathbf{NL}^{+}(\mathrm{k}\mathbf{v} - \mathbf{M}\boldsymbol{\alpha}) \tag{7.13}$$

These calculations look a little daunting at first, but the matrices **N, L**, and $\mathbf{L}^+$ have to be calculated only once for each experiment, so it is not an exorbitant amount of work to calculate $\boldsymbol{\beta}$ in this way for each analysis as we begin to treat the data post-experiment.

Case 3. Recalibration of the mass spectrometer using experimental data

If the calibration of the mass spectrometer were perfect and the measured peaks were noise free, then equation 7.11 would be an exact equality

$$\mathbf{M}\boldsymbol{\beta} = \mathrm{k}\,\mathbf{v} \tag{7.14}$$

If, on the other hand, the measured peaks **v** were noise free but the calibration was a bit off, then the vector $\mathbf{M}\boldsymbol{\beta}$ would not be proportional to **v**, but could be made so by an appropriate recalibration of **M** based on the previous calculations. If for each i = 1,..,m we multiply row i of **M** by the *recalibration factor*

$$\frac{\mathrm{k}\,\mathrm{v}_\mathrm{i}}{\sum_{\mathrm{i}=1}^{\mathrm{n}} \mathrm{M}_{\mathrm{i,j}}\beta_\mathrm{i}} \tag{7.15}$$

then equation 7.14 will hold exactly for the recalibrated **M**.

Since the measured peaks **v** will generally have some noise, a single set of recalibration factors computed as in equation 7.15 would not in fact produce a perfectly calibrated **M**. However, an average of several sets of recalibration factors calculated over a short time (always possible in TSR runs) should eliminate the noise quite effectively, and produce a better, in principle a correctly updated, calibration matrix **M**.

This recalibration will not actually reduce the amount of calculation needed, since one must still use equations 7.9, 7.12, and 7.13 to calculate each $\boldsymbol{\beta}$ in order to ensure that mass balance holds. In fact, frequent recalibration will add noticeably to the work, since for each recalibrated matrix **M** it will be necessary to recalculate the **L** and $\mathbf{L}^+$ matrices of equation 7.13. The main benefit of calculating the recalibration factors of equation 7.15 would be to track changes over time in the calibration of the mass spectrometer. Such changes could compensate for, or simply reveal, drift in the mass spectrometer that might need to be investigated. Additionally, of course, any individual set of recalibration factors that seemed particularly unusual might signal some kind of breakdown of the mass spectrometer system.

Case 4. Unmeasured or unexpected components in the reacting fluid

It may happen that the reactor outlet stream has components that are know to exist but, for whatever reason, are not being measured by the device used for analysis (mass spectrometer, gas chromatograph, etc.). A somewhat surprising but happy fact is that, within limits, mass balance considerations allow us to estimate the correct content of both the measured and the unmeasured components.

We first consider a calculational procedure appropriate for a mass spectrometer. A rather simpler procedure suffices if the measuring device gives partial pressures directly, and this simpler procedure is outlined later.

The case of the mass spectrometer

Suppose that some of the components of the output are not measured, say the last few. If we denote by $\boldsymbol{\beta}_1$ and $\boldsymbol{\beta}_2$ the vectors of measured and unmeasured components respectively, then we may partition the mole output vector $\boldsymbol{\beta}$ as

$$\boldsymbol{\beta} = \begin{bmatrix} \boldsymbol{\beta}_1 \\ \boldsymbol{\beta}_2 \end{bmatrix} \tag{7.16}$$

Correspondingly, **M, A,** $\boldsymbol{\alpha}$, and **c** may be partitioned as

$$\mathbf{M} = [\mathbf{M}_1 \mathbf{M}_2], \quad \mathbf{A} = [\mathbf{A}_1 \mathbf{A}_2], \quad \boldsymbol{\alpha} = \begin{bmatrix} \boldsymbol{\alpha}_1 \\ \boldsymbol{\alpha}_2 \end{bmatrix}, \quad \mathbf{c} = \begin{bmatrix} \mathbf{c}_1 \\ \mathbf{c}_2 \end{bmatrix} \tag{7.17}$$

The unmeasured part, $\mathbf{M}_2$, of the calibration matrix may be taken as all zeros. In fact, it does not enter into any of the calculations, and may be ignored. However, we must know the atomic structure of the unmeasured components given in $\mathbf{A}_2$, and the amount of any of these components present in the inlet stream, given by $\boldsymbol{\alpha}_2$, although this too may be all zeros. We also assume that the number of unmeasured components is small enough that that the columns of $\mathbf{A}_2$ are linearly independent, and independent of the columns of $\mathbf{A}_1$

Then the measured outlet mass spectrometer peaks **v** are given by:

$$\mathbf{M}_1\boldsymbol{\beta}_1 \cong \mathrm{k}\mathbf{v} \tag{7.18}$$

allowing for noise, and for an appropriate multiplier k. Now the mass balance equation 7.3 in partitioned form becomes

$$\mathbf{A}_1\boldsymbol{\alpha}_1 + \mathbf{A}_2\boldsymbol{\alpha}_2 = \mathbf{A}_1\boldsymbol{\beta}_1 + \mathbf{A}_2\boldsymbol{\beta}_2 \tag{7.19}$$

Denoting $\boldsymbol{\gamma}_2 = \boldsymbol{\beta}_2 - \boldsymbol{\alpha}_2$, the mass balance equation becomes

$$\mathbf{A}_1\boldsymbol{\alpha}_1 = \mathbf{A}_1\boldsymbol{\beta}_1 + \mathbf{A}_2\boldsymbol{\gamma}_2 \tag{7.20}$$

Just as in the previous procedure where all components are measured, we may calculate a constituent vector $\mathbf{c}_1$ that gives the least-squares approximate solution to

$$\mathbf{M}_1\mathbf{c}_1 \cong \mathbf{v} \tag{7.21}$$

namely

$$\mathbf{c}_1 = (\mathbf{M}_1^t\mathbf{M}_1)^{-1}\mathbf{M}_1^t\mathbf{v} \tag{7.22}$$

From equations 7.18 and 7.21 we have that

$$\boldsymbol{\beta}_1 \cong k\mathbf{c}_1$$

and substituting this into the mass balance equation 7.20 yields

$$\mathbf{A}_1\boldsymbol{\alpha}_1 \cong k\mathbf{A}_1\mathbf{c}_1 + \mathbf{A}_2\boldsymbol{\gamma}_2 = \left[\mathbf{A}_1\mathbf{c}_1\ \mathbf{A}_2\right]\begin{bmatrix} k \\ \boldsymbol{\gamma}_2 \end{bmatrix} = \mathbf{D}\begin{bmatrix} k \\ \boldsymbol{\gamma}_2 \end{bmatrix} \tag{7.23}$$

where $\mathbf{D} = [\mathbf{A}_1\mathbf{c}_1\ \mathbf{A}_2]$.

The best (least-squares) approximate solution to this is

$$\begin{bmatrix} k \\ \boldsymbol{\gamma}_2 \end{bmatrix} = (\mathbf{D}^t\mathbf{D})^{-1}\mathbf{D}^t\mathbf{A}_1\boldsymbol{\alpha}_1 \tag{7.24}$$

We note that the assumption at the beginning, that the columns of $\mathbf{A}_2$ are linearly independent and independent of the columns of $\mathbf{A}_1$, guarantees that the matrix $\mathbf{D}^t\mathbf{D}$ is invertible.

This gives estimates for k and γ_2 that can now be used to estimate $\boldsymbol{\beta}$. Given these estimated values of k and γ_2 we seek the least-squares approximate solution to equation. (7.18), subject to the linear constraint in equation 7.20. This solution turns out to be (see the mathematics section below):

$$\boldsymbol{\beta}_1 = \boldsymbol{\alpha}_1 - \mathbf{A}_1^+\mathbf{A}_2\boldsymbol{\gamma}_2 + \mathbf{N}\mathbf{L}^+(k\mathbf{v} - \mathbf{M}_1\boldsymbol{\alpha}_1 + \mathbf{M}_1\mathbf{A}_1^+\mathbf{A}_2\boldsymbol{\gamma}_2) \tag{7.25}$$

where $\mathbf{A}_1^+ = \mathbf{A}^t(\mathbf{A}_1\mathbf{A}_1^t)^{-1}$
$\mathbf{L} = \mathbf{M}_1\mathbf{N}$
$\mathbf{L}^+ = (\mathbf{L}^t\mathbf{L})^{-1}\mathbf{L}^t$

Also, by the definition of $\boldsymbol{\gamma}_2$, we get the unmeasured components of $\boldsymbol{\beta}$ as

$$\boldsymbol{\beta}_2 = \boldsymbol{\alpha}_2 + \boldsymbol{\gamma}_2 \tag{7.26}$$

Procedures for mass spectrometer data with unanalyzed components

1. Partition:

$$\mathbf{M} = [\mathbf{M}_1\mathbf{M}_2],\quad \mathbf{A} = [\mathbf{A}_1\mathbf{A}_2],\quad \boldsymbol{\alpha} = \begin{bmatrix} \boldsymbol{\alpha}_1 \\ \boldsymbol{\alpha}_2 \end{bmatrix},\quad \boldsymbol{\beta} = \begin{bmatrix} \boldsymbol{\beta}_1 \\ \boldsymbol{\beta}_2 \end{bmatrix}$$

corresponding to measured and unmeasured components ($\mathbf{M}_2$ is irrelevant, and can be ignored), and compute:

$$
\begin{aligned}
\mathbf{N} &= \text{null matrix for } \mathbf{A}_1 \\
\mathbf{L} &= \mathbf{M}_1\mathbf{N} \\
\mathbf{L}^+ &= (\mathbf{L}^t\mathbf{L})^{-1}\mathbf{L}^t \\
\mathbf{M}_1^+ &= (\mathbf{M}_1^t\mathbf{M}_1)^{-1}\mathbf{M}_1^t \\
\mathbf{A}_1^+ &= \mathbf{A}_1^t(\mathbf{A}_1\mathbf{A}_1^t)^{-1}
\end{aligned} \tag{7.27}
$$

2. For each data vector **v**, compute:

$$
\begin{aligned}
\mathbf{c}_1 &= \mathbf{M}_1^+\mathbf{v} \\
\mathbf{D} &= [\mathbf{A}_1\mathbf{c}_1 \ \mathbf{A}_2] \\
\mathbf{D}^+ &= (\mathbf{D}^t\mathbf{D})^{-1}\mathbf{D}^t \\
\begin{bmatrix} k \\ \boldsymbol{\gamma}_2 \end{bmatrix} &= \mathbf{D}^+\mathbf{A}_1\boldsymbol{\alpha}_1 \\
\boldsymbol{\beta}_1 &= \boldsymbol{\alpha}_1 - \mathbf{A}_1^+\mathbf{A}_2\boldsymbol{\gamma}_2 + \mathbf{N}\mathbf{L}^+(k\mathbf{v} - \mathbf{M}_1\boldsymbol{\alpha}_1 + \mathbf{M}_1\mathbf{A}_1^+\mathbf{A}_2\boldsymbol{\gamma}_2) \\
\boldsymbol{\beta}_2 &= \boldsymbol{\alpha}_2 + \boldsymbol{\gamma}_2
\end{aligned} \tag{7.28}
$$

Procedure for gas chromatograph data with unanalyzed components

When a gas chromatograph is used for composition measurements, estimates of the partial pressures are given directly. Or rather, estimates are made directly of a constituent vector **c**, that if properly normalized would give the partial pressures. If, on the other hand, some components of the outlet stream are unmeasured, then it is not immediately clear how to normalize **c**. In fact we may use a simplified version of the procedure described above to solve this problem for a mass spectrometer. Again we ignore the calibration matrix $\mathbf{M}_1$ and the measured peaks vector v (i.e. set $\mathbf{M}_1$ equal to the identity matrix in the formulas above), and obtain the following procedure. Partition:

$$
\mathbf{A} = [\mathbf{A}_1 \quad \mathbf{A}_2], \quad \boldsymbol{\alpha} = \begin{bmatrix} \boldsymbol{\alpha}_1 \\ \boldsymbol{\alpha}_2 \end{bmatrix}, \quad \boldsymbol{\beta} = \begin{bmatrix} \boldsymbol{\beta}_1 \\ \boldsymbol{\beta}_2 \end{bmatrix}
$$

corresponding to measured and unmeasured components, and compute

$$
\begin{aligned}
\mathbf{N} &= \text{null matrix for } \mathbf{A}_1 \\
\mathbf{N}^+ &= (\mathbf{N}^t\mathbf{N})^{-1}\mathbf{N}^t \\
\mathbf{A}_1^+ &= \mathbf{A}_1^t(\mathbf{A}_1\mathbf{A}_1^t)^{-1}
\end{aligned} \tag{7.29}
$$

For each data vector **c** compute

$$\begin{aligned}
\mathbf{D} &= [\mathbf{A}_1\mathbf{c}_1 \quad \mathbf{A}_2] \\
\mathbf{D}^+ &= (\mathbf{D}^t\mathbf{D})^{-1}\mathbf{D}^t \\
\begin{bmatrix} k \\ \gamma_1 \end{bmatrix} &= \mathbf{D}^+\mathbf{A}_1\boldsymbol{\alpha}_1 \\
\boldsymbol{\beta}_1 &= \boldsymbol{\alpha}_1 - \mathbf{A}_1^+\mathbf{A}_2\boldsymbol{\gamma}_2 + \mathbf{N}\mathbf{N}^+(k\mathbf{c} - \boldsymbol{\alpha}_1 + \mathbf{A}_1^+\mathbf{A}_2\boldsymbol{\gamma}_2) \\
\boldsymbol{\beta}_2 &= \boldsymbol{\alpha}_2 + \boldsymbol{\gamma}_2
\end{aligned} \tag{7.30}$$

The above two procedures are auxiliary to the conventional processes of identifying the full composition vector of a reactor output. They would be used only at the beginning of an experimental program, when the missing components would be identified, and then included as known in the vector of output components for the rest of the experiment. Nevertheless, once the unanalyzed components are known, one could save some trouble and money by neglecting their direct analysis (for example, because it requires a separate analytical facility, or takes too much time) relying instead on the methods outlined to calculate their contribution to the output composition.

Summary of mass and atomic balancing

The balancing procedures described above start with a raw output composition vector and seek to produce a corrected output composition vector whose molar contents and atomic balances are such that they present a solution closest to a perfect balance in each of the components. The adjustments made to the raw data will in principle remove the analytical error present in the raw data and will quickly alert the operator to the presence of systematic or unusual random errors in the analysis. The procedures used for this purpose are based on the self-evident proposition that mass and atomic balances must all close between the input and the output of the reactor.

Because of the plentiful data collected from a TSR experiment it is particularly worthwhile in this type of operation to adjust the output compositions of individual analyses by subjecting the raw data to this balancing procedure at each analysis. However, even individual analyses in conventional studies can be improved in this way. The adjusted data is better and more representative of the real composition of the products, although it will not be perfect and will need further examination. In the case of large assemblies of data, as is the case with the data from a TSR experiment, this balanced data is subsequently processed by yet other reliability-enhancement methods that are applicable only to large assemblies of data. These will be described below when we consider filtering the data.

These several possibilities of enhancing large data bases by the use of mathematical techniques, as contrasted to building better but more expensive apparata, offers significant opportunities for cost savings while improving the reliability of the data collected for kinetic investigations.

Calculating Fractional Conversion

Whether we apply the above balancing procedure or not, the starting data required for the next step is a vector of fractional yields and conversions of components in the output. Many instruments will report the analyzed output composition in terms of mol fractions (or partial pressures) directly. However, as we saw in the above discussion, this raw data vector usually contains a composition that is not in atomic balance. The balancing described above produces the vector **β** which is a mass- and atomic-balanced vector but whose components do not add up to one and do not represent fractional conversions at the output. To get to that vector we need to renormalize the **β** vector by dividing each of its components by the sum of the components of **β**. The new vector **β′** reports mol fractions at the output that are atomically balanced, but not the fractional conversions. We will call it a *corrected composition vector.*

The components of the *raw composition vector* or of the *corrected composition vector* **β′** are still not the quantities required for data fitting with candidate rate expressions. The corrected composition vector is to be preferred but we could use either it or the raw composition vector for the next step. Either vector must now undergo processing to calculate the *fraction converted* and *fractional yields* of products from the *mol fractions* reported in the composition vector. To see why and how we do this we need to look at the following logic.

The output composition vector for a TSR consists of mol fractions of all components present at the output. We define the following terminology:

n_i = moles of component i;
y_i = mol fraction of component i;
n_{total} = sum of moles of all components, including all diluents.

Each individual output mol fraction (y_i) is therefore:

$$y_{i,out} = \frac{n_{i,out}}{n_{total,out}} \tag{7.31}$$

It will be convenient in this section to discuss these mol fractions and their derivative fractions in terms of individual components of the output vector rather than dealing with the output vector as a unit.

Now, for the purpose of fitting rate expressions we require fractional conversion or fractional yield quantities (x_i), such as fractional conversion

$$n_{react,out} = \frac{n_{react,in} - n_{react,out}}{n_{react,in}} \tag{7.32a}$$

and fractional yield:

$$x_{i,out} = \frac{n_{i,out}}{n_{react,in}} \tag{7.32b}$$

In applying the above definitions one must also select a *base component*. In the case of monomolecular reactions this is the sole reactant itself, but in the case of bimolecular reactions one must chose either one of the two reactants to be the base component. The second reactant is then tied to the behaviour of the base component, as we will soon see.

Transformation from the raw mol-fraction analytical data to the requisite fractional conversion usually requires knowledge of the *volume expansion factor*, ε. In simple cases this factor is defined as:

$$\varepsilon = \frac{n_{\text{total, 100\%conversion}} - n_{\text{total, 0\%conversion}}}{n_{\text{total, 0\%conversion}}} \tag{7.33}$$

After a number of algebraic manipulations of the above equations one arrives at the conclusion that the relationships between mol fractions and the required fractional conversions in the output stream are as follows.

The fraction of unconverted feed is

$$x_{react,out} = \frac{y_{react,in} - y_{react,out}}{y_{react,in} + \varepsilon\, y_{react,out}} \tag{7.34}$$

the fraction of converted feed is

$$1 - x_{react,out} = \frac{(1+\varepsilon)y_{react,out}}{y_{react,in} + \varepsilon\, y_{react,out}} \tag{7.35}$$

and the fractional yield of each product is

$$x_{i,out} = \frac{(1+\varepsilon)y_{i,out}}{y_{react,in} + \varepsilon\, y_{react,out}} \tag{7.36}$$

The above transformations yield the conversion fractions x_i as required for the treatment of kinetic rate expressions. In cases where there is no volume expansion, the factor $\varepsilon = 0$ and the above relationships collapse to the less complicated forms:

$$x_{react,out} = \frac{y_{react,in} - y_{react,out}}{y_{react,in}} \tag{7.37}$$

and

$$1 - x_{react,out} = \frac{y_{react,out}}{y_{react,in}} \tag{7.38}$$

and

$$x_{i,out} = \frac{y_{i,out}}{y_{react,in}} \tag{7.39}$$

The above formulas offer a well-established way of adjusting the raw, or corrected, analytical data containing mol fractions at the outlet to yield the appropriate quantities for use in rate expressions, written in terms of fractions of mols converted. However, in temperature scanning experimentation, where temperature is not constant during the period of data collection, these procedures are inadequate unless ε is expressed as a function of temperature.

To get around this problem we will use another more versatile but less well-known method of calculating the x_i values. We will examine its merits but should always remember that, regardless of the method used for establishing the expansion factor in vapor phase reactions, a measure of volume expansion in the reactor is required before the output composition data reported by the analytical system can be used in rate expressions.

Determining Volume Expansion

In an ideal situation we might follow the outlet and inlet volumetric flow rates directly. In practice this is rarely possible in high temperature reactions of gaseous feeds. In Chapter 5 we discussed methods for calculating the volume expansion factor δ_V, one using the stoichiometric factor ε, and the other based on direct measurement of outlet flow rates. In the following section we give more detail about methods for calculating ε, and subsequently discuss methods for calculating δ_V. In doing this we will see how these quantities can be obtained more directly using stream composition measurements.

The price of this approach is that we rely on the accuracy of the analytical data to quantify volume expansion as well as the fractional conversions themselves, making these values doubly important in the rate calculations and adding to the collection of reasons for a meticulous cleaning-up of the raw analytical data using the methods that will be outlined in this chapter below.

First we discuss various methods for calculating ε, the conventional quantifier of volume expansion. In doing this we will come to understand the numerous difficulties associated with its use in TSR experiments.

Calculating ε

There is a problem if we use a single value of ε in processing TSR data from gas phase conversions if parallel reactions are present, and in all the reactor types if secondary reactions contribute to volume expansion. The definition of ε is that given in equation 7.33 and its value will vary with temperature. It will vary even more with conversion unless the reaction studied is not paralleled by other reactions and the products are stable at reaction conditions.

Equation 7.40 makes it clear that in cases where parallel or consecutive reactions are present the output volumetric flow rate V_{out} will not vary linearly with the fraction converted under temperature scanning conditions, or indeed in conventional isothermal operation.

$$V_{out} = V_{in}(1 + \varepsilon x_{react.,out}) \tag{7.40}$$

A better defined and more robust way of formulating ε than that presented in 7.33 is to write it as:

$$\varepsilon = \frac{\sum_{prod,i} Sc_i . Se_i - 1}{1 + r_{inert,in}} \tag{7.41}$$

where Sc_i is the stoichiometric coefficient for the given product;
Se_i is the selectivity for the product;
$r_{inert,in}$ is the ratio of mols of inert per mol of reactant in the feed.

The above formulation also yields a simple constant for a given input composition and stoichiometry as long as all the products are the result of parallel primary reactions and conversion is carried out at isothermal conditions. Problems in using this more elaborate formulation for TSR studies become obvious as soon as we consider the case where products are formed in the same set of parallel reactions but at different temperatures. The factor that causes difficulty in the case of a TSR is the variation in selectivity for the parallel processes as temperature is ramped.

When products appear by parallel reactions, the selectivities for their formation are almost certain to change with temperature since the several parallel reactions are almost sure to have different activation energies. To apply this formula in TSR operations it would be necessary to map out the selectivity behaviour of all the parallel processes as a function of temperature and to apply appropriate expansion factors at each temperature during the ramp in a TSR experiment. But, of course, the selectivity is not known *a priori*, even if we wished to apply this procedure.

In cases where *series* reactions contribute to volume expansion the situation becomes even more complex, and it would be necessary to have ε as a function of both temperature and conversion in order to carry out the appropriate transformations. This could only be done after the TSR experiment is completed and would require a thorough mapping of the behaviour of the reaction within the range of reaction conditions being studied. One would have to know the behaviour of the kinetics before carrying out the study.

A different possibility is to try to determine ε directly from measured concentrations as the TSR experiment is being carried out. If there are a number of parallel reactions producing various products i, then, since the total number of moles converted is

$$n_{moles\ converted} = x_{react,out}\, n_{react,in}$$

it follows that for each product we have

$$n_{i,out} = n_{moles\ converted} \cdot Sc_i . Se_i = x_{react,out} \cdot n_{react,in} \cdot Sc_i . Se_i \tag{7.42}$$

Therefore we can write an expression for the selectivity of each product at each instant during the TSR experiment:

$$
\begin{aligned}
Se_i &= \left[\frac{n_{i,out}}{n_{react,in}}\right]\left[\frac{1}{x_{react,out}}\right]\left[\frac{1}{Sc_i}\right] \\
&= \left[\frac{n_{i,out}}{n_{total,out}}\right]\left[\frac{n_{total,in}}{n_{react,in}}\right]\left[\frac{n_{total,out}}{n_{total,in}}\right]\left[\frac{1}{x_{react,out}}\right]\left[\frac{1}{Sc_i}\right] \\
&= y_{i,out}\left[\frac{1}{y_{react,in}}\right](1+\varepsilon\, x_{react.,out})\left[\frac{1}{x_{react,out}}\right]\left[\frac{1}{Sc_i}\right] \\
&= \left[\frac{(1+\varepsilon\, x_{react,out})}{y_{react,in}\, x_{react,out}}\right]\left[\frac{y_{i,out}}{Sc_i}\right]
\end{aligned}
\tag{7.43}
$$

Since a similar equation holds for each product i, and since the sum of the selectivities is 1 by definition, we have:

$$1 = \sum_i Se_i = \left(\frac{1+\varepsilon x_{react,out}}{y_{react,in}\, x_{react,out}}\right)\cdot\sum_i\left(\frac{y_{i,out}}{Sc_i}\right) \tag{7.44}$$

Thus

$$\frac{(1+\varepsilon\, x_{react,out})}{y_{react,in}\, x_{react,out}} = \frac{1}{\sum_i \frac{y_{i,out}}{Sc_i}} \tag{7.45}$$

and so the selectivity for a given product is, from equation 7.43,

$$Se_{prod} = \frac{y_{prod,out}}{Sc_{prod} \sum_{prod\, i}\left(\frac{y_{i,out}}{Sc_i}\right)} \tag{7.46}$$

This gives a formula for the instantaneous selectivities of parallel reactions in terms of known and measured quantities. From this we can calculate ε at each instant of the TSR experiment. We can develop these formulas further in order to get a direct formula for ε. Since

$$\frac{1}{1+r_{inert,in}} = \frac{1}{1+\frac{n_{inert,in}}{n_{react,in}}} = \frac{n_{react,in}}{n_{react,in}+n_{inert,in}} = y_{react,in} \tag{7.47}$$

it follows from equations 7.41 and 7.46 that

$$\varepsilon = \left(\sum_{product} Sc_{prod} \cdot Se_{prod} - 1 \right) y_{react,in}$$

$$= \left(\sum_{prod} \frac{y_{prod,out}}{\sum_i \frac{y_{i,out}}{Sc_i}} - 1 \right) y_{react,in} \tag{7.48}$$

$$= \left(\frac{\sum_i y_{i,out}}{\sum_i \frac{y_{i,out}}{Sc_i}} - 1 \right) y_{react,in}$$

This, finally, gives ε for parallel reactions in terms of known stoichiometric factors and measured concentrations. To get the value of ε we need nothing more than an analysis of the input and output streams. We note that in the special case of a single reaction it reduces to the intuitively obvious formula

$$\varepsilon = (Sc_{prod} - 1)\, y_{react,in} \tag{7.49}$$

which in the case of pure reactant at the input gives $\varepsilon = (Sc_{prod} - 1)$.

Some difficulties in calculating ε

The formulas above, although valid for all parallel reactions, pose difficulties for a TS-PFR experiment. Because of varying conditions along the length of a PFR, selectivities may well also vary along the length. These formulas, however, are based on measured conditions that will be available only at the outlet of the reactor. They will therefore produce some sort of average ε, based on some sort of average selectivities along the reactor axis. One might hope that in a given TS-PFR experiment the reactor is operating at nearly isothermal conditions, i.e. with only a small temperature profile at any given clock time during the experiment. This gives hope that the selectivities will in fact be relatively constant through the reactor at any given clock time. Unfortunately, this approximation is not sure, and therefore should be avoided.

Still more complications arise if there are parallel reactions that produce certain products by more than one path. The concentration measurements y_i will not distinguish among these, so formula 7.48 will be indeterminate. There is a solution to some of these problems. The convenient method of calculating ε that avoids the above difficulties involves exploiting the presence of inerts in the reacting fluid, a well known method but one worth restating.

Using inerts to calculate ε

It is almost always possible to introduce an *inert*, or *tracer*, into the reacting fluid. In order to cause the least amount of dilution of the reactants, the type of inert used should be chosen to give consistently accurate measurements of its mol fraction at low mol fractions of the inert. With the inert mol fraction accurately measured, it is easy to get ε directly from the current reactor outlet composition. Since

$$n_{total,out} = n_{total,in}\left(1 + \varepsilon x_{react,out}\right) \tag{7.50}$$

and

$$\begin{aligned} 1 + \varepsilon x_{react,out} &= 1 + \varepsilon\left(\frac{y_{react,in} - y_{react,out}}{y_{react,in} + \varepsilon y_{react,out}}\right) \\ &= \frac{(1 + \varepsilon)x_{react,out}}{y_{react,in} + \varepsilon y_{react,out}} \end{aligned} \tag{7.51}$$

therefore, since $n_{inert,in} = n_{inert,out}$, we get:

$$\begin{aligned} y_{inert,in} &= \frac{n_{inert,in}}{n_{total,in}} \\ &= \left(\frac{n_{inert,out}}{n_{total,out}}\right)\left(\frac{n_{total,out}}{n_{total,in}}\right) \\ &= y_{inert,out}\left(1 + \varepsilon x_{react,out}\right) \\ &= y_{inert,out}\left(\frac{(1+\varepsilon)y_{react,in}}{y_{react,in} + \varepsilon y_{react,out}}\right) \end{aligned} \tag{7.52}$$

This may be solved algebraically for ε to give:

$$\varepsilon = \frac{1 - \dfrac{y_{inert,out}}{y_{inert,in}}}{\dfrac{y_{inert,out}}{y_{inert,in}} - \dfrac{y_{react,out}}{y_{react,in}}} \tag{7.53}$$

Since $y_{inert,in}$ and $y_{react,in}$ need to be measured only once throughout a TSR experiment, it is only necessary to track $y_{inert,out}$ and $y_{react,out}$ in order to calculate ε. The value of ε may well vary during a run due to changing temperature and perhaps conversion, but that does not affect the formula above. In particular, since the selectivities do not enter into this formula, it does not matter if they vary through the length of the reactor.

Formula 7.53 gives a very simple and satisfactory method for calculating the instantaneous value of ε whenever it is needed. However, once the virtue of using inerts for this calculation has been noted, the question arises whether ε is the best method of accounting for volume expansion, and hence if it is needed at all. The traditional uses for ε are in formulas for conversion fractions, as in equations 7.34 – 7.36 above, and in the reaction rate expression where terms like (1 + εx) are introduced to account for stoichiometric volume expansion (see for example O. Levenspiel (1999) third edition, page 68.) We shall soon see that all these cases can be handled just as easily by a more direct calculation of a different volume expansion factor, using an inert as a tracer or without it. The alternative procedure we believe to be more versatile and simpler to use in TSR applications.

Using inerts to calculate volume expansion and fractional conversion

Let A be the base reactant. Then the fractional conversion of a reactant A is defined, regardless of the chemistry, as

$$x_a = \frac{n_{a,in} - n_{a,out}}{n_{a,in}} = 1 - \frac{n_{a,out}}{n_{a,in}} \tag{7.54}$$

By inserting extraneous terms, this becomes

$$\begin{aligned} x_a &= 1 - \left(\frac{n_{a,out}}{n_{total,out}}\right)\left(\frac{n_{total,out}}{n_{inert}}\right)\left(\frac{n_{inert}}{n_{total,in}}\right)\left(\frac{n_{total,in}}{n_{a,in}}\right) \\ &= 1 - y_{a,out}\left(\frac{1}{y_{inert,out}}\right) y_{inert,in}\left(\frac{1}{y_{a,in}}\right) \\ &= 1 - \left(\frac{y_{a,out}}{y_{a,in}}\right)\left(\frac{y_{inert,in}}{y_{inert,out}}\right) \end{aligned} \tag{7.55}$$

This gives conversion of A as a simple expression involving only the inlet and outlet mol fractions of A and the inert.

Denote by δ the stoichiometric expansion factor $\delta = n_{total,out}/n_{total,in}$. This equals the previously defined volumetric expansion factor δ_V = (volume out)/(volume in) if there is no expansion due to heating or pressure change. In cases where there is expansion due to temperature or pressure changes, $\delta_V = \delta(TP_0/T_0P)$.

It follows that δ may also be written in terms of mol fractions, as

$$\begin{aligned} \delta &= \frac{n_{total,out}}{n_{total,in}} = \frac{\left(\dfrac{n_{inert}}{n_{total,in}}\right)}{\left(\dfrac{n_{inert}}{n_{total,out}}\right)} \\ &= \frac{y_{inert,in}}{y_{inert,out}} \end{aligned} \tag{7.56}$$

Substituting the result in terms of output mol fractions into the previous expression for x_a gives:

$$x_a = 1 - \delta \frac{y_{a,out}}{y_{a,in}} \tag{7.57}$$

All the above equations, written in terms of δ, hold regardless of the network of reactions present, including consecutive reactions or selectivity changes due to temperature ramping. They follow simply from the definitions of fractional conversion and stream composition.

For products, the fraction converted to a given product is defined as:

$$x_{prod} = \frac{n_{prod}}{n_{a,in}} \tag{7.58}$$

Thus

$$x_{prod} = \frac{n_{prod}}{n_{a,in}} = \frac{\left(\frac{n_{prod}}{n_{total,out}}\right) n_{total,out}}{\left(\frac{n_{a,in}}{n_{total,in}}\right) n_{total,in}} = \delta \frac{y_{prod,out}}{y_{a,in}} \tag{7.59}$$

We finally arrive at the following short collection of simple formulas for expansion δ and conversion x that can be used anywhere that ε has traditionally been used. They avoid all questions about changing selectivities, and in fact totally avoid all questions about the stoichiometry of the reactions. Moreover δ can be easily calculated for each analysis of a TSR experiment, making it possible to update volume expansion effects along the temperature ramp of each TSR run of an experiment. In TSR work the following formulas are therefore the preferred relationships for calculating fractional quantities for kinetic fitting.

$$\delta = \frac{y_{inert,in}}{y_{inert,out}}$$

$$x = 1 - \delta \frac{y_{react,out}}{y_{react,in}} \tag{7.60}$$

$$x = \delta \frac{y_{prod,out}}{y_{react,in}}$$

The above convert the raw data, available in terms of mol fractions from the analytical equipment or after balancing to give the corrected mol fraction vector calculated in the manner described above, to fractional yields and conversions used in rate expressions, as required.

The Relationships among Composition Vectors

At this point we need to review and clarify the relationships between the composition vectors **α, β** and **p** and include in our considerations the vector containing the corrected output mol fractions $y_{i,out.}$ The vector **α** contains the mol fractions of components in the feed. The components of the vector **α** obviously add up to 1. The components of this vector are also the $y_{i,in}$ quantities appearing in equation 7.60 and before. This vector contains all the molecular species present in the output composition but the elements not present in the feed are set to zero.

The vector **β**, on the other hand, is a vector of the number of moles present in the output stream *per mol of total input*. Its components need not add up to 1. A moment's reflection shows that the sum of the components of **β** is the number of moles in the output stream per mol of total input and hence represents the volume expansion factor δ. Using **β** one can generate the vector of output mol fractions $y_{i,out}$ by dividing each component of the **β** vector by the sum of its components, renormalizing **β** so its components add up to 1. In this way we obtain the corrected mol fractions $y_{i,out}$ for use in equations 7.60 and before. However, the simplest procedure for using equations 7.60 and the like is to set δ =1 and use the components of **β** in place of the corresponding $y_{i,out}$, without bothering to calculate the $y_{i,out}$ themselves.

This does not mean that the **β** vector is never used in any other way. Calculating the value of δ and the $y_{i,out}$ is still necessary for subsequent model fitting and simulation. Moreover, calculating the vector of corrected $y_{i,out}$ values is useful in that comparison of the corrected $y_{i,out}$ vector with the corresponding raw data **p** vector of the output composition tells us much about the validity of the analysis in that instance.

The vector **β** is a very useful and little appreciated quantity, giving the volume expansion factor δ and allowing a simplified calculation of the fractional yields and conversions $x_{i,j}$ required for kinetic fitting. Notice also that this vector is available for each analysis. It can therefore be used in treating even single-point results such as those obtained in standard tests of catalysts, or elsewhere in the study of conversions. The calculation of the **β** vector and the corresponding corrected quantities of the output **y** vector should be a routine procedure in treating all analytical results.

> *By calculating β we immediately obtain the expansion factor δ, and ε if we should need it, without the necessity of introducing an inert diluent or tracer into the reacting mixture. We can also readily see whether the raw analytical results, such as those composing the vector **p** in equation 7.4, differ significantly from the corrected $y_{i,out}$ vector calculated from β. If they do, there is reason to suspect the accuracy of the analysis and/or infer the presence of missing, unanalyzed, components.*

Review of Traditional Rate Expressions Involving Volume Expansion.

One of the traditional uses for ε is in rate expressions where terms like (1+εx) are introduced to account for stoichiometric volume expansion. These cases can be easily handled by the above method of calculation of volume expansion using the presence of inerts in the feed to calculate ε. For example, in Levenspiel (1999), pp. 69-72, it is shown that the rate for a variable volume reaction involving reactant A can be written as:

$$-r_A = \frac{C_{A0}}{1+\varepsilon_A x_A}\frac{dx_A}{dt} \tag{7.61}$$

Similarly the rate expressions for first and second order reactions can be written as:

$$-r_A = \frac{kC_{A0}(1-x_A)}{1+\varepsilon_A x_A} \tag{7.62}$$

and

$$-r_A = kC_{A0}^2\left[\frac{(1-x_A)}{1+\varepsilon_A x_A}\right]^2 \tag{7.63}$$

Levenspiel correctly observes that these expressions depend on volume expansion being linear with conversion. This is a simplification which is not valid in temperature scanning operations and is merely an approximation in most practical situations. Now, since $(1+\varepsilon_A x_A)$ is just the volume expansion factor δ, and since this factor can be updated at each analysis in TSR operations, the above rate expressions may alternatively be written as:

$$-r_A = \frac{C_{A0}}{\delta}\frac{dx_A}{dt} \tag{7.64}$$

$$-r_A = \frac{kC_{A0}(1-x_A)}{\delta} \tag{7.65}$$

$$-r_A = kC_{A0}^2\left[\frac{(1-x_A)}{\delta}\right]^2 \tag{7.66}$$

Similarly easy revisions apply to other forms of rate expressions involving volume expansion. These formulas do not require any assumption about linearity of volume expansion with conversion, nor are they invalidated by the presence of parallel or consecutive reactions. One simply needs to establish the instantaneous values of δ at each analysis, a matter made simple if inerts are present in the feed or by calculating δ from the β vector.

Other reactants may be present in the inlet feed, as for example in the case of bimolecular reactions. Their fractional conversion may be treated the same way as the base reactant, or may be treated the same way as reaction products. The first of these is probably the preferred approach. These two approaches will obviously give rise to different rate expressions when expressed in terms of fractional conversions but will be the same when rate expressions are expressed in terms of concentrations. Indeed, we see from the fact that $y_{react,in}$ is another name for C_{A0}, that in a rate expression a typical term of the form $C_{A0}(1-x_A)/\delta$ is simply equal to the outlet concentration $y_{react,out}$, and that a term of the form $C_{A0}x_{prod}/\delta$ is equal to the outlet concentration $y_{prod,out}$.

Similarly, even if there is already some product present in the inlet stream, the fractional conversion of the product may be handled as for any other product. This comes about because $x_{prod,out}$ is simply an alternate way of expressing the outlet concentration of the product.

Difficulties with Expansion in the Case of Noisy Data.

One potential difficulty in using equations 7.60 arises when the raw analytical data used to calculate the various quantities is noisy and needs to be smoothed. This, unfortunately, occurs in almost every real TSR experimental situation. Powerful techniques for smoothing the data exist and are discussed in detail in subsequent sections. At the moment, the difficulty we need to examine is that both $x_{react,out}$ and δ involve the use of mol fractions from the analysis of the output stream and that $x_{react,out}$ therefore is affected by errors in the raw data twice over, amplifying any error present in the raw data. The question is: at what point should the smoothing be done.

If $x_{react,out}$ is to be smoothed this will affect the values of $y_{react,out}$. This in turn affects the values of $y_{inert,out}$. In turn this affects the value of δ, which has been used in the calculation of the un-smoothed values of the various $x_{react,out}$ via equation 7.60. This feedback effect would seem to require some iterative process to reach a final, consistent, smoothed value of $x_{react,out}$. A simpler way around the difficulty is to consider a slightly revised version of equations 7.60, one where $x_{react,out}$ is not explicitly dependent on δ. For instance, equation 7.65 can be written as:

$$x_{react,out} = 1 - \left(\frac{y_{inert,in}}{y_{react,in}}\right)\left(\frac{y_{react,out}}{y_{inert,out}}\right) \tag{7.67}$$

In this form, we see that smoothing $x_{react,out}$ can be thought of as, in effect, adjusting the values of $y_{react,out}$ and $y_{inert,out}$ so as to give a smooth behaviour in $x_{react,out}$. The constant values of $y_{inert,in}$ and $y_{react,in}$ can be supposed to have been carefully measured once and for all, and so they need not be adjusted and represent the best analytical result possible. This will, as before, affect the value of δ, but now there is no feedback to $x_{react,out}$, so no iterative re-adjustments are needed. Similarly, equation 7.59 can be written without explicit dependence on δ as:

$$x_{prod,out} = \left(\frac{y_{inert,in}}{y_{react,in}}\right)\left(\frac{y_{prod,out}}{y_{inert,out}}\right) \tag{7.68}$$

There still remains the question of smoothing δ. This must be dealt with since δ will be needed for the revised forms of rate expressions we will be using. There are two different methods one might adopt: one very simple and satisfactory for most cases, and the other rather more complicated but useful in cases where large smoothing corrections are needed.

The simple procedure is to ignore the smoothing effect of $x_{react,out}$ and $x_{prod,out}$ on the $y_{i,out}$ values and simply smooth δ separately, after the corrected vectors have been calculated, allowing the $y_{i,out}$ to assume different values. The smoothing is over the range of conditions that were investigated and produces an appropriate value of smoothed δ for each analysis. The rationale for this is that we may consider *modest* smoothing of, say, $x_{react,out}$, as moving it toward its true (noiseless) value. Similarly, *modest* smoothing of δ moves it toward its true value. Any slight residual error due to not having considered possible interactions of the two smoothings should not be significant if the smoothing required is minor. Moreover, it is not necessary for the smoothing of δ to be as thorough as that for x_{react}, where sensitive numerical derivatives must be calculated. All in all, this uncoupled smoothing method should be satisfactory if the noise level of the data is low.

The more complicated procedure is to take account of the effect on δ of smoothing $y_{react,out}$ and $y_{prod,out}$. To do this, rewrite equations 7.67 and 7.68 in terms of output mole fractions, as they might be reported by some analytic instrument.

Since $y_{react} = n_{react,out}/n_{total,out}$, and similar expressions hold for other concentrations, equations 7.67 and 7.68 may be written respectively as:

$$x_{react,out} = 1 - \left(\frac{y_{inert,in}}{y_{react,in}}\right)\left(\frac{n_{react,out}}{n_{inert,out}}\right) \tag{7.69}$$

$$x_{prod,out} = \left(\frac{y_{inert,in}}{y_{react,in}}\right)\left(\frac{n_{prod,out}}{n_{inert,out}}\right) \tag{7.70}$$

Suppose, as we may if the inert was well chosen for analytical sensitivity, that the measurements of $n_{inert,out}$ are accurate. In that case any changes in $x_{react,out}$ and $x_{prod,out}$ due to smoothing are the result solely of changes in $n_{react,out}$ and $n_{prod,out}$. From the preceding equations the relation between these changes is seen to be

$$\Delta n_{react,out} = -n_{inert,out}\left(\frac{y_{react,in}}{y_{inert,in}}\right)\Delta x_{react,out} \tag{7.71}$$

and

$$\Delta n_{prod,out} = n_{inert,out}\left(\frac{y_{react,in}}{y_{inert,in}}\right)\Delta x_{prod,out} \tag{7.72}$$

where the Δx_i values are the difference between corresponding values of the smoothed and the un-smoothed data sets. From this a new value of $n_{total,\,out}$ can be calculated by adding $\Delta n_{total,\,out}$ to the old value to get:

$$n_{total,out}^{revised} = n_{total,out} + \Delta n_{total,out}$$
$$= n_{total,out} + n_{total,out}\left(\frac{n_{inert,out}}{n_{total,out}}\right)\left(\frac{y_{react,in}}{y_{inert,in}}\right)(\Delta x_{prod,out} - \Delta x_{react,out}) \tag{7.73}$$
$$= n_{total,out}\left[1 + y_{inert,out}\left(\frac{y_{react,in}}{y_{inert,in}}\right)(\Delta x_{prod,out} - \Delta x_{react,out})\right]$$

Now one can calculate a revised value of $y_{inert,out}$ as:

$$y_{inert,out}^{revised} = \frac{n_{inert,out}}{n_{inert,out}^{revised}} = \frac{y_{inert,out}}{\left[1 + y_{inert,out}\left(\frac{y_{react,in}}{y_{inert,in}}\right)(\Delta x_{prod,out} - \Delta x_{react,out})\right]} \tag{7.74}$$

Finally, from this new value of $y_{inert,out}^{revised}$ one can calculate a revised value of $\delta = y_{inert,in} / y_{inert,out}^{revised}$. If the $x_{react,out}$ and the $x_{prod,out}$ have been appropriately smoothed, this will result automatically in a smoothed value of δ.

The above discussion points out an important feature of temperature scanning methods. The collected data is amenable to sophisticated methods of error removal. The availability of these methods in temperature scanning comes from the large amounts of data gathered and the internal consistencies that can be expected to exist in the data set from an experiment.

The removal of the random and relatively small errors present as noise in TSR data is accomplished by means of filtering, a topic that receives more attention below as we consider fitting rate expressions using TSR data. The above procedures calculate quantities used in rate expressions but the rates themselves have to come from slopes of smooth curves.

Dealing with Noise in Experimental Data

In TSR experiments, as in any physical system, the readings taken will be subject to a certain amount of inaccuracy. Numerous techniques are available to reduce the effects of this random error, traditionally involving averaging methods of one sort or another, e.g. arithmetic averaging at a single point, piecewise approximations of the data by simple functions, or curve fitting of the total data collected. In the analysis of data from TS-PFR experiments, more than in the case of most data analysis, it is critically important to employ an appropriate error reduction technique, since in the extraction of rates from the raw TS-PFR data it is usually necessary to find the *slopes* of various surfaces. Averaging over several readings is not possible because there are no repeat readings at any point, and curve fitting will not do since it prejudges the shape of the curve. What *is* possible is smoothing of the data using filtering algorithms.

Zero-Dimensional Data Smoothing

The most familiar form of data smoothing is the process of averaging a series of readings at a point. The procedure usually applied is *arithmetic averaging*, a procedure of dividing the sum of several readings by the number of readings being summed. Other averages, such the *logarithmic average*, where the sum of logarithms of the readings is divided by their number, may be more appropriate in some cases, but are not often used.

More elaborate averages involve weighting of individual readings to change their contribution to the sum, so that the more reliable readings, or some region of the variable space, weigh more heavily on the final average. The weighting factors are determined from considerations such as run conditions or reading-instrument properties, but most often are limited to outright rejection of "wild" points, i.e. putting a weight of zero on such points and leaving them out of the sum, based on intuition or more substantive information available to the data interpreter.

Systematic error

The averages obtained by these several procedures are almost certain to be more reliable than a single reading but do not even attempt to eliminate another type of error: systematic error. This type of error shifts all readings away from the true value because of instrumental or calibration error. None of the methods discussed here are capable of eliminating systematic error. Nor is there any mathematical procedure capable of revealing its presence.

Systematic error is the hidden danger in all experimentation and must be eliminated by extraneous means, such as calibration and verification of readings before beginning the experimental program. Often one can also spot systematic error by applying independent crosschecks appropriate to the situation. When results are published, readers of the report have to take it as an article of faith that systematic error has been eliminated from the data. They have the right to expect that the data reported is free of systematic distortions. Remember also that systematic error can appear in *any* of the dimensions of a data point.

One-dimensional Data Smoothing

In one-dimensional smoothing we are concerned with reducing error in both the position of a curve and its shape. It is an unfortunate and well known fact that not every good approximation to the *position* of a curve will yield a good approximation to the *shape* of the same curve, as represented by its slopes.

Smoothing by moving windows

Consider, for example, the function $f(x) = \exp(x)$ on the interval [1,2], whose slope is also exp(x). This graph is shown in as a solid line in Figure 7.1. To it has been added random error that we will call *noise*. In our example the noise is normally distributed with mean 0 (i.e. no systematic error) and standard deviation 0.25. We examine how one might set about identifying and eliminating this noise in a way appropriate to TS-PFR data.

A simple smoothing procedure is to average each value with the two adjacent values. This form of smoothing is said to use a *moving window* of length 3. We first use a window with equal weights of 1/3 for each point in the triplet. Figure 7.1(a) shows that this averaging has brought the points significantly closer to the original curve. However, when one estimates slopes by taking symmetric differences at each pair of points, the results, shown in Figure 7.1(b) are not very encouraging.

The approximation to the curve can be improved somewhat by using more elaborate "windowing". For instance, Figure 7.2 shows averaging using a window of length 5 with triangular weights 1/9, 2/9, 3/9, 2/9, 1/9.

Figure 7.3 shows averaging using a window of effective length 5 with Gaussian weights. In this case the weights given to the y_j are proportional to $K(t)=(x_i - x_j)/2$ where $K(t) = 1/0.927 \exp(-t^2/0.274)$. This Gaussian averaging method, along with other sophisticated filtering techniques, can be found in commonly available commercial mathematical packages, such as MathCad. In this example, however, none of these methods improves the derivative estimates in any significant way. Since there is no rational way of deriving a good estimate of the smooth behaviour of the first derivative from slope data such as that shown in Figures 7.1b, 7.2b and 7.3b, one must look for a filtering method that achieves satisfactory smoothing of the function itself *and* of its first derivative, *simultaneously*.

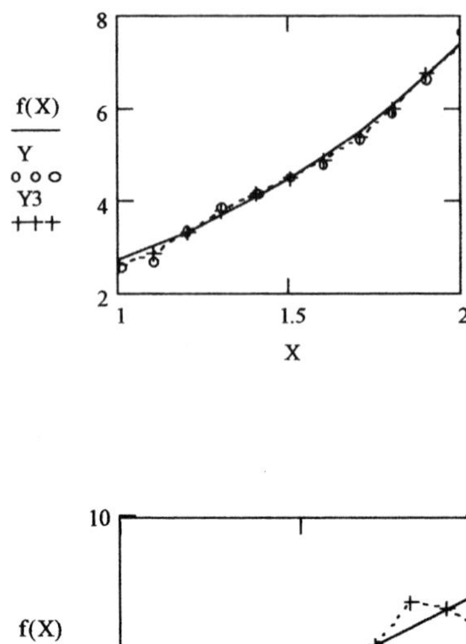

Figure 7.1(a).
Circles show the noisy data about the true curve. Crosses show moving averages using a window of length three.

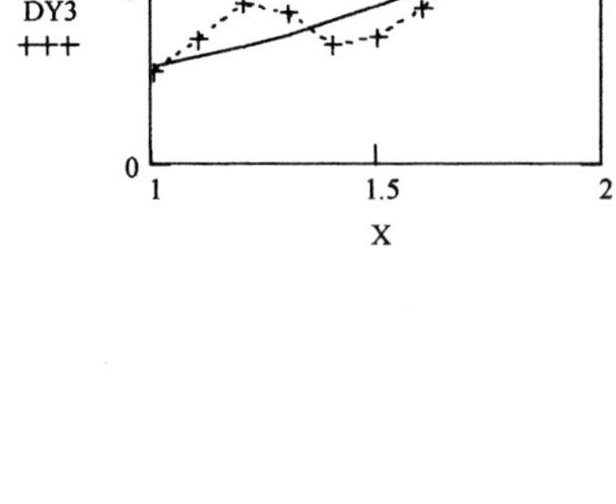

Figure 7.1(b).
True and estimated slopes using symmetric differences.

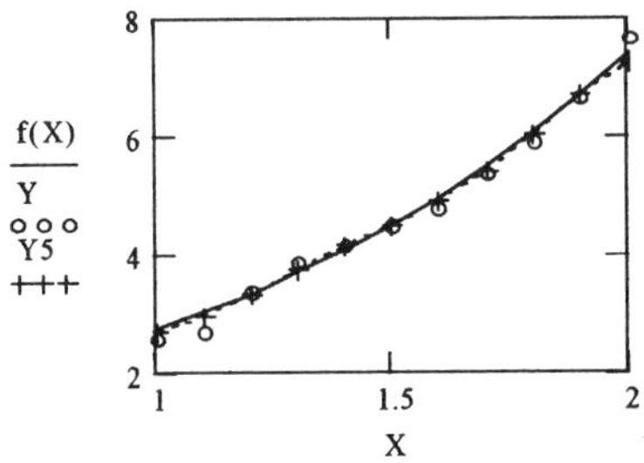

Figure 7.2(a)
Crosses show the moving average with a window of length five and weights in ratio 1,2,3,2,1.

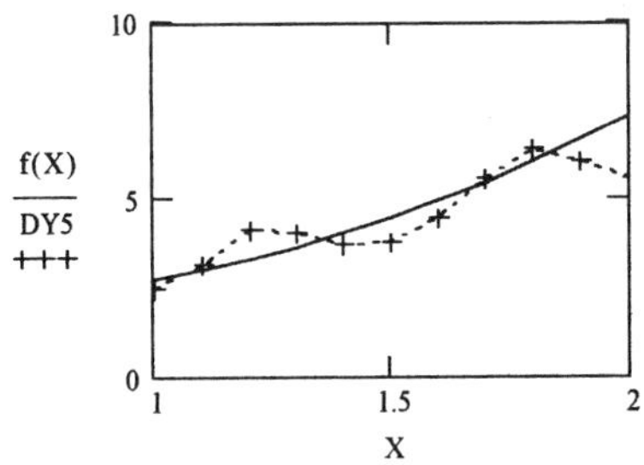

Figure 7.2(b).
True and estimated slopes using symmetric differences.

Figure 7.3(a).
Crosses show the moving average with a window of length 5 and Gaussian weights.

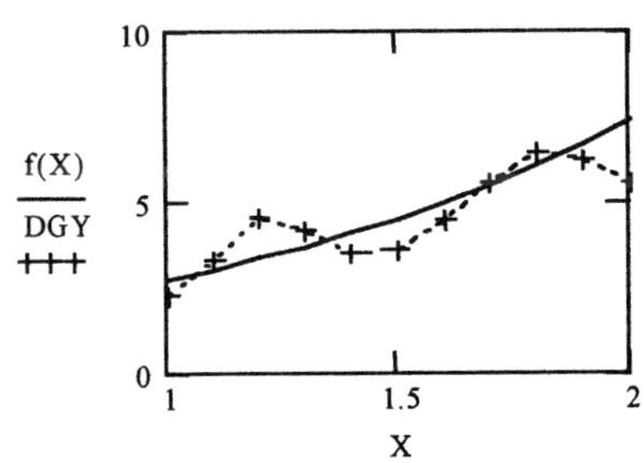

Figure 7.3(b)
Estimated slopes using symmetric differences from filtered data in Figure 7.3a

In the case of TSR data, and in our example, a much more successful approach to establishing both the position and the shape of the curve lying behind the error-containing data is to approximate the curve piecewise, using simple smooth functions, usually low-order polynomials. The derivatives of these approximating functions then give estimates of the derivatives of the desired curve. One such method is that of moving least squares (see Davis (1973 and 1975)). An even better technique, one that has proved quite successful in the treatment of TSR data, is that of spline functions.

Smoothing by splines

To obtain smooth curves of derivatives of noisy data, such as can arise from a TSR experiment, one wants an approximation, rather than an exact interpolation, that will closely match both the curve and its derivative. *Splines*, and especially *cubic splines*, provide such algorithms and have long been used for interpolation. They are now a standard component of many data analysis packages. The standard reference is that by de Boor (1978). Cubic splines used for interpolation of one-dimensional data are piecewise cubic functions that are chosen to pass through the selected data points and also have their pieces matched together so that the spline has a continuous derivative. *B-splines* have proven to be most useful for the treatment of TSR data. B-splines (or basic splines) come in all orders. It will suffice here to consider only B-splines based on quadratic polynomials. This is not to say that further work may not reveal that other splines have an advantage in treating TSR data. Work on applications to this kind of data is just starting and improvements are almost certain to be made.

Given any four points (knots) on the x-axis, a B-spline for these points consists of three quadratic pieces joined together at the knots and chosen so that the spline has a continuous derivative and vanishes outside the interval containing the points. De Boor (1972) gives an efficient and stable method of calculating such B-splines. On an interval containing the experimental data, one may choose some appropriate number of knots, including three additional knots artificially placed at each end of the interval to give initial and final B-splines, in appropriate positions, and form the B-splines for each quartet of knots in turn. For instance, on the interval [1,2] one may choose internal knots at, say, 1.3 and 1.65. We then get the five B-splines shown in Figure 7.4 (normalized so that at each point the sum of the B-splines is 1.

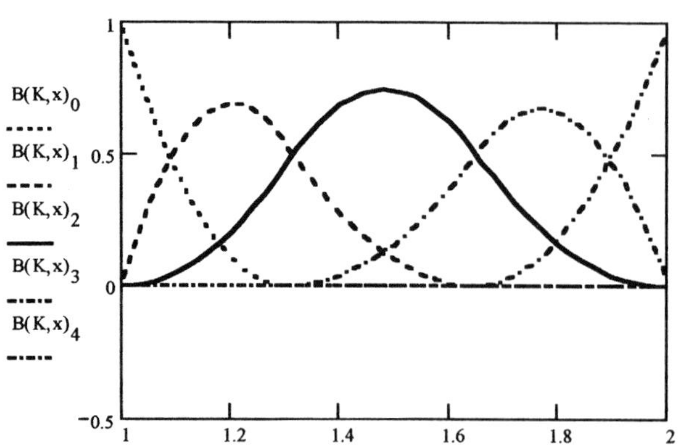

Figure 7.4.
The five B-splines on [1,2], with internal knots at 1.3 and 1.65.

Given some such B-splines, one seeks a linear combination of them that best fits the data. The first thought is to find the linear combination that minimizes the sum of square differences between the spline and the data. By itself, however, this criterion neglects the desire that the spline should also give a smooth approximation to the derivative of the curve that generated the data. To achieve this, we also want the spline to be a smooth as possible. To satisfy this requirement we employ another strategy.

The B-spline has a continuous first derivative because we require this but it has discontinuities in the second derivative. Another desideratum that we can therefore employ is that the sum of squares of these second-derivative discontinuities be as small as possible. It turns out that the twin goals of fitting the data and maximizing the smoothness of the first derivative in this way compete with each other. This is a problem that will be with us in all higher dimensions.

> *It requires judgment, based on some broader understanding of the expected results, to find the appropriate balance between the competing requirements in data fitting in any particular situation. A purposeful search for a defensible optimum solution, rather than an objective impartial minimum, is the goal.*

Figures 7.5 and 7.6 show two cases of fitting the above B-splines to the data shown in Figures 7.1, 7.2, 7.3 (i.e. on the interval [1,2], exp(x) + noise). In Figure 7.5 the coefficients in the linear combination are chosen to give a reasonable balance between fitting the position and getting a smooth spline. As can be seen, the derivative estimates are remarkably good using this criterion.

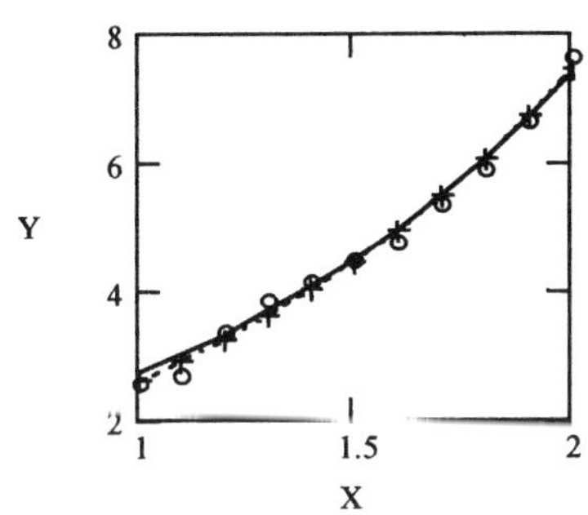

Figure 7.5(a).
Crosses show the spline, balanced between fitting position and smoothness.

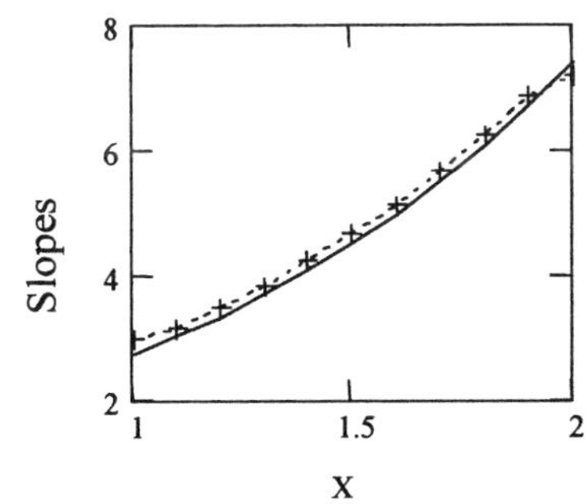

Figure 7.5(b).
Estimated slopes spline in 7.5a show a good fit to the derivative of the curve.

In contrast to this, in Figure 7.6 the coefficients in the linear combination are chosen simply to give the best possible fit to the location of the data. This single criterion of fit does not serve well. In this case the spline is not nearly as smooth and the derivatives oscillate considerably. As a result they give poor estimates of the slopes, as poor as the averaging techniques we abandoned above.

The smoothing of one-dimensional data is therefore far from being as mechanical, or indeed as objective, as the commonly used un-weighted arithmetic averaging of zero-dimensional data. There are numerous considerations involved in curve fitting by B-splines. How many knots are needed? What is their best placement? Iterative methods that attempt to make the choices more objective have been developed for these procedures (Dierckx 1982), and a package of FORTRAN routines implementing them (called DIERCKX) is available on NETLIB at http://www.netlib.org/dierckx/. Similar smoothing algorithms are also available in the commercial statistical analysis program S-Plus.

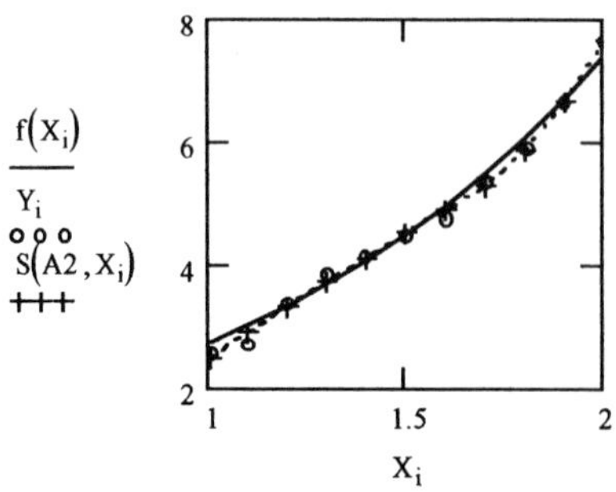

Figure 7.6(a).
Crosses show the spline chosen simply to best fit the position of the data.

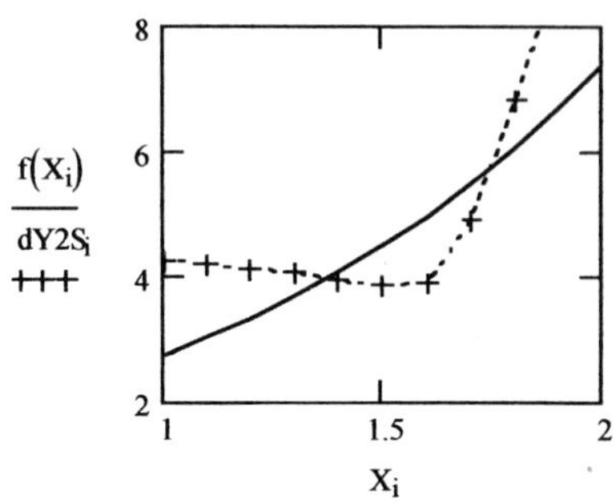

Figure 7.6(b).
Estimated slope using the derivative of the fitted spline in Figure 7.6(a). Note the poor fit to the derivative of the curve

The application of smoothing procedures involving B-splines, and other such techniques recommended for dealing with one- and two-dimensional data, is to some significant extent dependent on the needs of the user. The methods most appropriate to kinetic data have yet to be defined. TS-PFR data smoothing procedures are not as mechanical in their application as procedures used in treating error in zero-dimension data. In fact, they are distinct enough from traditional averaging procedures that their application to TSR data to obtain the best possible estimates of the curves, and the slopes underlying the data, constitutes a special type of error correction technique called *data filtering* and the many sophisticated procedures available for this purposes are referred to as *filters*.

The correct application of filters is central to the success of a TS-PFR experiment since it is the slopes obtained by this filtering that are the rates of the reaction. It would be less than wise to apply outdated or inappropriate methods of error handling to data obtained by a new technique that offers so much promise and such different data.

Smoothing by filtering

Finally, we mention another filtering technique that may be used when there is a lot of data collected at regular time intervals, such as one might have in a TSR experiment. For such data it is possible to employ filtering techniques similar to those used in digital communications. The essential idea in this case is to remove various select frequency components from the data by transforming the data to the frequency domain, cutting off aberrant frequency components, then transforming the filtered data back to the time domain. There are many different algorithms for achieving this, and programs implementing them are widely available. For a data set as small as the examples above this is not an appropriate technique, but for the wealth of data produced by a TS-PFR experiment it could be competitive with spline fitting, in the case of one-dimensional data. The unresolved issue is whether TS-PFR data, after conversion to the frequency domain, contains frequencies where the noise generated by experimental error is concentrated. There is some evidence for this, given in a different context by Bonanzi (1996), but as yet no direct test has been attempted with TS-PFR data. The answer may well depend on the method used to measure the conversion.

It may also be that in the future several filtering techniques will be applied in sequence to the same data-set, as various aspects of the noise are treated in succession. There is a need to define such procedures and understand their applications. Many new procedures and much understanding of reactor system behaviour will come from such work.

Two-Dimensional Data

One might think that the one dimensional data from each run in a TS-PFR experiment should be filtered using one of the above methods and that would clean up the pertinent errors. This is not so. The data arising from a set of runs in a TS-PFR experiment need to be treated as a two-dimensional data-set, since in order to extract reaction rates it is necessary to find slopes between points *from different runs*. What needs to be done in this case is the following: data from the whole TS-PFR experiment must be assembled into a *kinetic surface* and the surface smoothed as a whole. Slopes taken from this surface, in specific directions, are the rates of reaction.

Because of inevitable noise in the data, the data kinetic surface, even after the mass and atomic balancing described above, will be rough, and we must filter this data before slopes are taken if we are to avoid magnified noise in the slopes (rate data). This issue was illustrated above for the one dimensional case but with the TS-PFR we are dealing with noise in two dimensions. To obtain slopes with a low noise level from TS-PFR data we need to obtain a smooth 2-dimensional surface (draped in 3-dimensional space) that is formed from a limited number of noisy one-dimensional traverses of the surface, the *runs* of the TS-PFR *experiment*. It will not do to filter the traverses one by one. This would leave unfiltered the discrepancies *between* the several runs in an experiment. We must filter all the data from an experiment as a single data-set as we prepare it for rate extraction. From this smooth surface of experimental conversion data we will obtain a smooth surface of its first derivatives – the surface of reaction rates.

As in the one-dimensional case, spline functions can be used. For instance, Hayes and Halliday (1974) describe an efficient least-squares method for fitting bi-cubic splines to data arranged arbitrarily. A bi-cubic spline is a piecewise cubic in two variables whose pieces fit together in quadrilateral "patches" to give a function of two variables with continuous partial derivatives. However, the algorithms described do not take into account the important question of how many knots should be used to determine the grid on which the cubic spline pieces fit or where they should be placed. A procedure that automatically chooses the number and placing of the knots is a two-dimensional generalization of the algorithms mentioned above due to Dierckx (1982). It is part of the DIERCKX package available at NETLIB. This filter uses two-dimensional B-splines that may be obtained as follows.

If M(x) is a B-spline based on some knots x_i on the x-axis, and N(y) is a B-spline based on some knots y_j on the y-axis, then M(x)N(y) is a B-spline based on the rectangle of knots (x_i, y_j); i.e. it is a function in two variables, with continuous partial derivatives, that vanishes outside its rectangle of knots. As in the one-dimensional case, one seeks linear combinations of these B-splines such that they fit both the data position as well as possible, and make the spline surface as smooth as possible. The relative weight to put on these two competing criteria is a parameter that the operator must supply, based on experience with the particular situation and the nature of the data. In practice, one might well try different relative weights and examine the results visually to decide which parameter choice gives results that look best. Judgment is therefore even more an issue in the filtering of two-dimensional data.

Another advantage of the B-spline filter is that there are very fast and accurate recurrence schemes for evaluating and differentiating the resultant surfaces (de Boor, (1972); Cox (1972)). This means that once a good surface of experimental data has been determined from a TS-PFR experiment, there is a fast process for calculating rates at *any* desired point on that surface within the range of the experimental data. This capability will produce the *rate surface* and is the source of the unlimited number of rates available from TS-PFR experiments.

What these procedures do is ensure that, when the smoothed data is fitted to a candidate rate expression, we are fitting the information contained in the whole kinetic surface, rather than that from a few isolated points. Here, then, is the definitive advantage of TSR techniques. Rather than seeking to fit a rate expression to a small set of error-containing points, as is normally done in dealing with data from conventional experimentation, we are attempting to reproduce the shape of a well defined kinetic surface. In doing this we can place emphasis on fitting particularly well any portion of the surface, without repeating the experiment. Alternatively, we can pursue a variety of criteria of "best" fit to suit our requirements. Moreover, we can re-filter the data in various ways before fitting the rate expression in order to examine the validity of our procedures and conclusions. As a result, data interpretation in TSR experimentation becomes more challenging and informative. It also offers the opportunity to validate kinetic and mechanistic conclusions in unprecedented detail, *with little need of further experimentation.*

Data Fitting

It is now clear how the application of filtering to TS-PFR data produces a smooth X-τ-T surface from which the dX/dτ slopes, that constitute the reaction rates, can be obtained at any point. Figure 7.7 shows the resultant *conversion surface* for a simple exothermic reversible reaction. Temperature increases left to right, conversion is on the vertical axis and space time increases front to back. The ridge of maxima is characteristic of reversible exothermic reactions and occurs where the temperature reaches a point where the reverse reaction begins to reduce product yield.

Slopes taken along certain highly specific directions on this surface yield the rates of reaction. The specific directions of these slopes are easily identified because on this surface we can also identify the *operating lines* that pertain to this experiment. Operating lines were discussed in Chapter 3 and we need to recall that these are traces of readings that would have been taken in successive ports of a long reactor operating in the heat transfer regime of the actual TS-PFR used. In order to obtain valid rates we therefore need to find these operating lines on the conversion surface and identify the sets of data points that lie along them.

Conversion vs Exit Temperature for Various Space Times

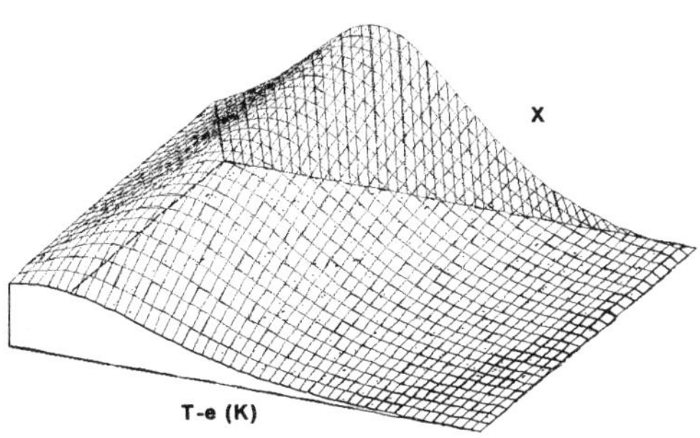

Figure 7.7
Conversion surface plotted on the temperature-tau plane. Slopes from this plane, taken in the direction of the operating lines, give the rates of reaction at all points of the surface.

Of all the possible slopes at each point on the conversion surface, it is the slope taken along the operating line through that point (and there is only one operating line through each point) that yields the correct rate of reaction. By taking rates at each point on the conversion surface we obtain a corresponding point-by-point mapping of rates on the same space-time-temperature plane. More importantly these rates can be mapped onto a plane of conversion vs. temperature. This is what we call an experimentally determined *kinetic surface*. The shape of this surface is defined by a kinetic rate expression:

$$\text{rate} = f(\text{conversion}(X), \text{temperature}(T)).$$

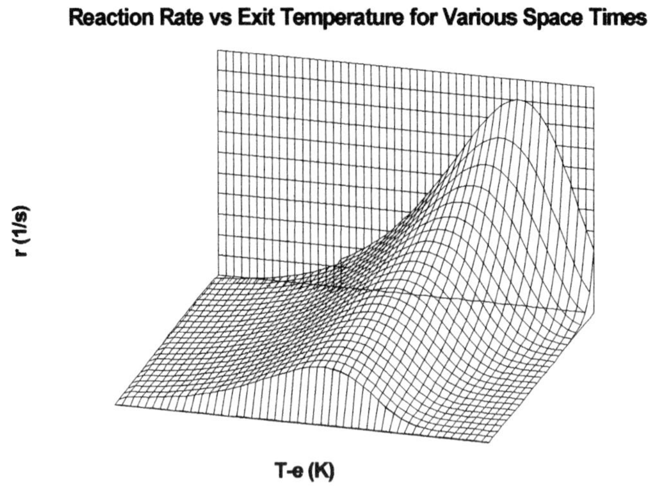

Figure 7.8
Kinetic or rate surface obtained from slopes taken from the conversion surface shown in Figure 7.7. The analytical rate expression, be it mechanistic or empirical, is an algebraic function that reproduces the shape and location of this surface for a given reaction at specified reaction conditions.

The exercise of fitting a candidate rate expression to TSR experimental data, and finding the appropriate rate parameters, is therefore one in which we seek an analytical form of the rate expression that, with appropriate parameters, fits the shape and position of the rate surface over the entire range of the experimental conditions used. This is a much more demanding and satisfactory criterion than the fitting of a complex rate expression to the few isolated points normally available from conventional kinetic studies.

Notice that all this mapping and slope-taking can be done using readily available computer routines. The procedure yields the desired rates and other specifics of each data point without the necessity of integrating the defining equations presented in Chapter 5. This not only avoids potentially messy mathematical manipulations but makes data interpretation fairly straightforward and easy to program on a computer. Figure 7.8 shows a map of a kinetic surface for a reversible exothermic reaction that may be encountered in practice.

Along the front axis in Figure 7.8 we plot the reaction temperature increasing from left to right. Front to back is increasing space time. The height represents the net rate of reaction. We see that as the reaction moves to higher temperatures at constant space time, at first the rate increases. It reaches a maximum and begins to decrease as the reverse reaction of the exothermic equilibrium begins to gain prominence. The maximum rate achieved decreases from back to front as space time increases at constant temperature. The differences between the kinetics of similar exothermic reactions is contained in the shape and location of very similar surfaces, each specific for that reaction. Those are the differences that let us distinguish between rival rate expressions.

The surface shown in Figure 7.8 is more extensive than that normally approachable by experiment. In most cases only a part of this surface can be investigated experimentally, say a part of the up-sloping area to the left of the ridge. Information about the difference between rival rate models is then contained in the positioning and curvature of just a fragment of the total picture. No wonder that the few isolated readings on this surface made available by conventional means are quite inadequate to identify the correct rate expression with any degree of certainty. As we will see, even the masses of data made available by the TSR are highly stressed to identify a unique rate expression in such cases. The three dimensional view of kinetic surfaces makes plain the need for the collection of extensive rate data before a unique, hopefully mechanistic, rate expression can be identified. Rates that lie below the ridge and well to the left lie in a fairly flat region; and the data may not contain enough information to limit the range of possible rate expressions to a significant extent. This is especially true if the available data in this region consists of a sparse set of isothermal points with error attached, as is currently common in this type of work. Hence the persistent uncertainties in the interpretation of much kinetic work done to date.

What one requires for a good kinetic fit is not just a well defined curvature to differentiate the available surface-area from a flat plane or other similar shapes, but a kinetic surface free of roughness or noise. It also helps a great deal if information can be obtained about the behaviour of the surface near its ridge of maximum rates, or at other topographical "features" of more complex surfaces. Since the range of accessible experimental conditions may preclude investigations near the ridge, it becomes even more critical to define, precisely, the shape and position of as much as possible of the kinetic surface that may lie within the range of accessible experimental conditions.

Filtering of TS-PFR data to obtain the required smooth yet undistorted kinetic surface is a new skill for kineticists, every bit as important to the TS-PFR user as careful experimental work. Skillful data treatment will present the user of a TS-PFR with smooth surfaces in many sets of co-ordinates. This makes it possible to begin to understand the reaction before any attempt is made to fit a rate expression. For example, the interpreter of TSR data has the opportunity to examine and interpret the experimentally determined kinetic surface qualitatively, from visual inspection, even before an analytical fit is attempted. Such inspection can reveal topographical, and by implication mechanistic, features of special interest to industry or the theoretician.

It may well be that, with the availability of large data storage capacity, kinetic surfaces will become the data base of choice on which reactor design and control will be based. Users of this digital data need never resort to an analytical interpretation involving a rate expression.

This would leave the formulation of rate expressions to those workers who are interested in understanding the mechanism if the reaction. Many other users of TSR kinetic data would be adequately served by the various three-dimensional kinetic surfaces made available from filtered data obtained from a single TS-PFR experiment.

Suggested Procedures for Data Clean-up

The various methods of error removal should be carried out in some routine way so that data is treated systematically, avoiding random distortions that might be introduced by the clean-up procedures themselves. The order of operations might be as follows:

1. apply the atomic balancing algorithm;
2. calculate volume expansion;
3. calculate conversions and yields
4. examine volume expansion as a function of conversion and note behaviour.

If there is a disturbing lack of smoothness when procedure 4 is applied, re-examine the data for causes.

5. If the noise in procedure 4 is acceptable, go on to a one dimensional filtering of the conversion data.

Examine all the data to see if the traces of individual runs are congruous. If so;

6. Proceed with the two dimensional filtering by various two dimensional filters.

This will finally produce a surface suitable for the extraction of rate data.

There are several steps at which parity plots, described in Chapter 10, should be used to verify that no distortion was caused by the procedure just applied. Parity plots should be applied at as many of the intermediate steps in the clean-up as possible, always using the raw data as the base for comparison, although parity plots between steps in the clean-up can also be informative. If there is any reason to believe that systematic distortion is being introduced at some stage, the criteria used in that part of the clean-up should be changed to reduce the scale of the effect. After several different sequences of clean-up procedures, the differences between correctly cleaned-up sets of data should be negligible.

A Quick Review of Matrix Operations in Mass Balancing

Least-Squares Solutions Subject to Linear Constraints

In this section we give a brief review of the mathematics involved in solving the linear least-squares problems, with or without linear constraints, as encountered in this chapter. More details about the unconstrained problem can be found in any text on linear regression, e.g. Draper and Smith (1981), or Press, et al. (1986). More details about the constrained problem can be found in Lawson and Hanson (1974).

Unconstrained Least-Squares

The goal in the unconstrained least squares problem, given some $n \times m$ matrix $\mathbf{M}$ and an n-component vector $\mathbf{y}$, is to find an m-component vector $\mathbf{x}$ for which $\mathbf{Mx}$ is as close as possible to **y** in the least squares sense, i.e. such that the vector norm $\|\mathbf{Mx} - \mathbf{y}\|$ is as small as possible. In component form this can be written as follows. Find $x_1, x_2, \ldots x_m$ to minimise

$$S = \sum_{i=1}^{n}\left(\sum_{j=1}^{m} m_{ij}x_j - y_i\right)^2 \tag{7.75}$$

Differentiating with respect to each x_k in turn, we see that the at the minimum we must have, for each k,

$$0 = 2\sum_{i=1}^{n}\left(\sum_{j=1}^{m} m_{ij}x_j - y_i\right)m_{ik} = 2\left[\sum_{i=1}^{n}\sum_{j=1}^{m} m_{ij}m_{ik}x_j - \sum_{i=1}^{n} y_i m_{ik}\right] \tag{7.76}$$

With the order of summation reversed, these become the *normal equations*

$$\sum_{j=1}^{m}\sum_{i=1}^{n} m_{ij}m_{ik}x_j = \sum_{i=1}^{n} m_{ik}y_i \tag{7.77}$$

and in matrix form, these become the *matrix normal equation*

$$\mathbf{M}^t\mathbf{M}\mathbf{x} = \mathbf{M}^t\mathbf{y} \tag{7.78}$$

In most cases, and in particular in the cases we are dealing with in atomic balancing, the columns of the matrix **M** are linearly independent, and hence the matrix $\mathbf{M}^t\mathbf{M}$ is invertible. This finally yields the formula used in several places in the balancing sections. The least squares solution to $\mathbf{Mx} \cong \mathbf{y}$ is:

$$\mathbf{x} = (\mathbf{M}^t\mathbf{M})^{-1}\mathbf{M}^t\mathbf{y} \tag{7.79}$$

This formula, with various different notations, gives equations 7.6, 7.9, 7.12, 7.22, and 7.21. Equations 7.6 and 7.12 look different, but in fact they do follow from equation 7.79 since, $(\mathbf{Ap})^t(\mathbf{Ap})$ e.g., is a scalar, so that

$$[(\mathbf{Ap})^t(\mathbf{Ap})]^{-1} = \frac{1}{(\mathbf{Ap})^t(\mathbf{Ap})}$$

In equations 7.13, 7.25, and 7.27 we have denoted $\mathbf{L}^+ = (\mathbf{L}^t\mathbf{L})^{-1}\mathbf{L}^t$, and similarly for $\mathbf{M}_1^+$ in equation 7.27, and $\mathbf{D}^+$ in equation 7.28, and $\mathbf{N}^+$ in equation 7.29. This notation is used because these are all special cases of the Moore-Penrose Pseudoinverse $\mathbf{M}^+$ which can be defined for an arbitrary matrix **M** and which gives the minimum least-squares approximation even in cases where the columns of **M** may not be linearly independent (see Lawson and Hanson, 1974). Similarly, in equations 7.25, 7.27, and 7.29 we have denoted $\mathbf{A}_1^+ = \mathbf{A}_1^t\,(\mathbf{A}_1\mathbf{A}_1^t)^{-1}$ since this is another special case of the Moore-Penrose Pseudoinverse, for the case where the matrix in question, $\mathbf{A}_1$, has linearly independent rows.

Constrained Least-Squares

In the previous sections we several times sought the least-squares approximate solution to some linear equation subject to some linear constraint. In other words, we sought the best approximate solution to equation 7.4 subject to the constraint equation 7.3. Given **kp**, find $\boldsymbol{\beta}$ to solve

$$\boldsymbol{\beta} \cong \mathbf{kp}, \text{ subject to } \mathbf{A}\boldsymbol{\beta} = \mathbf{A}\boldsymbol{\alpha} \tag{7.80}$$

and for equation 7.11, subject to equation 7.3, given **kv**, find $\boldsymbol{\beta}$ to solve

$$\mathrm{M}\boldsymbol{\beta} \cong \mathbf{kv}, \text{ subject to } \mathbf{A}\boldsymbol{\beta} = \mathbf{A}\boldsymbol{\alpha} \tag{7.81}$$

For equation 7.18, subject to equation 7.20, given **kv**, find $\boldsymbol{\beta}$ to solve

$$\mathbf{M}_1 \boldsymbol{\beta}_1 \cong \mathbf{kv}, \text{ subject to } \mathrm{A}1\beta1 = \mathrm{A}1\alpha1 - \mathrm{A}2\,\gamma2 \tag{7.82}$$

These problems are all of the same general form: find the least squares approximate solution to

$$\mathbf{Mx} \cong \mathbf{y}, \text{ subject to } \mathbf{Ax} = \mathbf{b} \tag{7.83}$$

To solve this general problem we may proceed as follows. First, find a *null matrix* for **A**, i.e. a matrix **N** such that every solution to the homogeneous equation $\mathbf{Ax} = \mathbf{0}$ is given uniquely as $\mathbf{x} = \mathbf{Nz}$ for some **z**. Procedures to find such a null matrix are given in most elementary linear algebra books (see, for example, Lay, 1993), and are a standard application in, for example, the commercial software package Matlab. A computer-oriented algorithm for **N** that will handle numerically difficult cases is given in Press et al. (1986).

Now, given a null matrix **N**, and any particular vector $\mathbf{x}_0$ satisfying $\mathbf{Ax}_0 = \mathbf{b}$, it follows that every solution to $\mathbf{Ax} = \mathbf{b}$ is of the form

$$\mathbf{x} = \mathbf{x}_0 + \mathrm{Nz} \tag{7.84}$$

Substituting this into equation 7.83 gives

$$\mathbf{Mx}_0 + \mathbf{MNz} \cong \mathbf{y} \tag{7.85}$$

Where $\mathbf{L} = \mathbf{MN}$, this may be written:

$$\mathbf{Lz} \cong \mathbf{y} - \mathbf{Mx}_0 \tag{7.86}$$

There is no restriction on **z**, so the least-squares solution to this unconstrained problem is, following equation 7.86,

$$\mathbf{z} = \mathbf{L}^+(\mathbf{y} - \mathbf{Mx}_0) \tag{7.87}$$

where $\mathbf{L}^+ = (\mathbf{L}^t\mathbf{L})^{-1}\mathbf{L}^t$. Substituting this into equation 7.85 gives this final result. The least squares solution to equation 7.83, $\mathbf{Mx} \cong \mathbf{y}$, subject to $\mathbf{Ax} = \mathbf{b}$ is given by

$$\mathbf{x} = \mathbf{x}_0 + \mathbf{NL}^+(\mathbf{y} - \mathbf{Mx}_0) \tag{7.88}$$

where $\mathbf{x}_0$ is any particular solution to $\mathbf{Ax}_0 = \mathbf{b}$
$\mathbf{N}$ is a null matrix for $\mathbf{A}$
$\mathbf{L}$ $= \mathbf{MN}$
$\mathbf{L}^+$ $= (\mathbf{L}^t\mathbf{L})^{-1}\mathbf{L}^t$

This general result may now be applied to each of the three particular problems we considered, shown in equations 7.81, 7.82, and 7.83. For the problem in equation 7.81 we observe that the coefficient matrix $\mathbf{M}$ is the identity, and that $\boldsymbol{\alpha}$ may be taken as a particular solution to the constraint equation. We thus have $\mathbf{L} = \mathbf{N}$, and hence the general solution in equation 7.87 reduces immediately to the solution given in equation 7.7.

For the problem shown in equation 7.82 we again may take $\boldsymbol{\alpha}$as a particular solution to the constraint equation. The general solution in equation 7.87 then reduces to the solution given in equation 7.13.

For the problem shown in equation 7.83 it is not immediately apparent how to find a particular solution to the constraint equation. However, since the rows of $\mathbf{A}_1$ are assumed to be linearly independent, the matrix $(\mathbf{A}_1\mathbf{A}_1^T)$ is invertible. Thus we may take as a particular solution

$$\mathbf{x}_0 = \boldsymbol{\alpha}_1 - \mathbf{A}_1^+\mathbf{A}_2\boldsymbol{\gamma}_2 \tag{7.89}$$

where $\mathbf{A}_1^+ = \mathbf{A}_1^t(\mathbf{A}_1\mathbf{A}_1^t)^{-1}$, and verify that $\mathbf{A}_1\mathbf{A}_1^+ = \mathbf{I}$ so that $\mathbf{A}_1\mathbf{x}_0 = \mathbf{A}_1\boldsymbol{\alpha}_1 - \mathbf{A}_2\boldsymbol{\gamma}_2$ as required. With this value of $\mathbf{x}_0$, the general solution in equation 7.87 reduces to the solution given in equation 7.25.

Algorithms for Finding N, a Matrix Giving the Null Space of A

Take a simple example of the matrix $\mathbf{A}$ with rows giving the atomic components and columns the molecular components.

$$\begin{matrix} 2 & 0 & 0 & 1 & 0 & 2 \\ 4 & 0 & 0 & 0 & 2 & 4 \\ 0 & 2 & 0 & 2 & 1 & 1 \\ 0 & 0 & 1 & 0 & 0 & 0 \end{matrix}$$

In many such cases we can find $\mathbf{N}$ by partitioning $\mathbf{A} = [\mathbf{A}_1\mathbf{A}_2]$, with $\mathbf{A}_1$ a square invertible matrix, and taking:

$$\mathbf{N} = \begin{bmatrix} -\mathbf{A}_1^{-1} & \mathbf{A}_2 \\ & \mathbf{I} \end{bmatrix}$$

For the matrix **A** in the example above, this yields

$$\mathbf{N} = \begin{bmatrix} -0.5 & -1 \\ -1.5 & -0.5 \\ 0 & 0 \\ 1 & 0 \\ 1 & 0 \\ 0 & 1 \end{bmatrix}$$

If, however, the first columns of $\mathbf{A}_1$ are not linearly independent, one needs to adjust the **A** matrix. The easiest way is to change the ordering of the columns of **A** so that $\mathbf{A}_1$ is invertible. This amounts to rearranging the order in which the molecular components are listed. It is easy to do this by hand for any particular reaction - just start by listing the molecular components in an appropriate order so that an invertible $\mathbf{A}_1$ is produced. A user who is aware of these problems can avoid them by writing the column and row components in an order designed to avoid singularities, and then proceeding as above. In writing a permanent data interpretation program, however, one does not want to lay this extra burden on the programmer or the user. The necessary adjustments in the indices of a data interpretation program would be very confusing. It is preferable to use a general purpose algorithm for finding matrix **N**.

We note that the matrix **A** is normally small with integer entries, so that a simple Gaussian reduction algorithm should do the job of inversion. Numerical Recipes and Matlab both offer rather sophisticated algorithms, based on Singular Value Decomposition. These can be used although they are designed for large matrices. The fact that using these procedures requires more calculation than is necessary does not matter in practice, since **N** has to be calculated only once in each case of data treatment.

8. Fitting Rate Expressions to TSR Data

Hard science is the process of summarizing observable facts in the form of laws and formulas. These are often based on postulates of causal relationships that arise from the operation of mechanisms whose details are not observed directly. Progress consists of proposing alterations in these postulated mechanisms so as to enhance the validity, applicability and generality of the resultant laws and formulas.

Fitting Rate Expressions to Experimental Data

In Chapter 4 we observed that fitting rate expressions to experimental data can be a very difficult task. The usual approach is to find parameters that minimize the sum of squares (SS) of differences between experimentally observed values and fitted values. This process is repeated for each candidate rate expression and the solution with the lowest sum of squares is chosen as the best fit solution. However, the rate parameter estimates are often highly correlated, and the correlation is usually very non-linear. As a result, the numerical SS minimization problem is generally very ill conditioned, and therefore confidence in the fitted parameter values is small. In Chapter 4 some techniques were described that help alleviate these difficulties by employing measures such as temperature centering and the transformation of the rate expression. Although these methods help, the problem continues to be serious. In the end it turns out that the solution of these difficulties is going to be highly dependent on the availability of large amounts of high quality data.

The only method of improving the fits and our confidence in the estimates of kinetic parameters is therefore to find a way of increasing the amount of *good* experimental data for the fitting of rate expressions, made available in a reasonable amount of time and at reasonable cost. Conventional experimentation has not been able to do this in reasonable time and at tolerable cost. A breakthrough allowing the acquisition of a large volume of data was needed. This is where TSR methods come in: the advantage of temperature scanning reactors lies precisely in the amount of experimental rate data they provide. In Chapter 7 it was shown how a smooth rate surface can be constructed to provide rate data for every point in the X-T plane within the region traversed by the experimental runs. This yields two important benefits.

1. Data chosen for model fitting is plentiful and evenly spread over the whole X-T region investigated. In principle this can include all the space approachable with that reactor. This ensures that the fitting of the rate expression with be guided by a well documented shape to the rate surface.
2. Any amount of additional data may be obtained in specific regions of interest (like folds or regions of high curvature, etc.) without doing additional experimental work. This can be used to generate additional weight and ensure that the fitted rate expression has the correct behaviour in "topologically" interesting regions.

This is roughly equivalent to using a weighted least-squares in the fitting of sparse conventional sets of data. It has the advantage over the weighting procedure in that it uses more actual data points rather than simply putting an arbitrary extra weight on a few, possibly sparse, possibly erroneous, points in the region of interest.

> *Both these benefits address the heart of the rate expression fitting problem: the difficulty of identifying a unique rate expression and establishing the best set of its rate parameters.*

Optimization Algorithms

Once a smooth rate surface has been determined by processing experimental data, and a suitable rate expression has been chosen, we need to estimate the parameters of this rate expression by applying an optimization routine to solve the least-squares fitting problem. Many different optimization procedures are available, with no single procedure clearly better than all the others. The NEOS Guide to Optimization Software[2] lists approximately eighty different optimization software packages for non-linear minimization, both commercial and non-commercial packages. Perhaps new ones will be constructed specifically to fit kinetic data. In this section we briefly outline some of the most commonly used techniques, and mention their inclusion in several widely used general software packages. More details about the various algorithms can be found in Fletcher (1987), Luenberger (1984) and Press (1986).

The Problem

The basic problem can be stated as follows: minimize an *objective function* $f(\mathbf{x})$, where $\mathbf{x}$ is a vector variable comprising the parameters to be estimated, and $f(\mathbf{x})$ is the sum of squares to be minimized. For instance, in the oxidation of carbon monoxide, already mentioned in Chapter 4 and discussed in more detail in Chapter 11 (and in [Asprey, 1997]), the following Langmuir-Hinshelwood Dual Site Dissociative Adsorption Model is used to fit experimental data:

$$r_{CO} = \frac{k_r K_a K_b^{1/2} P_{CO} P_{O_2}^{1/2}}{1 + K_a P_{CO} + K_b^{1/2} P_{O_2}^{1/2}} \tag{8.1}$$

$$\text{where}\quad k_r = A_r e^{-E_r/RT}$$

$$K_a = A_a e^{E_a/RT}$$

$$K_b = A_b e^{E_b/RT}$$

In this case the parameter vector $\mathbf{x}$ has six components: $\mathbf{x} = (A_r, E_r, A_a, E_a, A_b, E_b)$. We may think of r_{CO} as a function of 8 variables. If we denote by $r(P_{CO}, P_{O2})$, the rate values actually obtained experimentally (i.e. values from the smoothed rate surface described

---

[2] (http://wwwfp.mcs.anl.gov/otc/Guide/SoftwareGuide/index.html)

in Chapter 7) for various partial pressures P_{CO}, P_{O2}, then our goal in fitting data is to find the parameters **x** that minimize the sum of squares:

$$f(\mathbf{x}) = \sum_{P_{CO}, P_{O2}} \left[r_{CO}(\mathbf{x}, P_{CO}, P_{O2}) - r(P_{CO}, P_{O2})\right]^2 \tag{8.2}$$

Information Required to Apply Various Optimization Algorithms

A few minimization algorithms work directly with the function values $f(\mathbf{x})$. Most, however, also use derivative values for the function being fitted, either first derivatives or second derivatives. The derivatives may be provided by the operator, or calculated analytically by the fitting program, or estimated numerically by the program. These issues have important implications for the applicability and facility of data fitting methods. To clarify these matters we need to understand the requirements of several of the most popular fitting methods, and how they influence the choice of fitting methods for TSR data. Denote:

g(x) = gradient vector of first derivatives

$$= \nabla f(\mathbf{x}) = \left(\frac{\partial f}{\partial x_1}, \frac{\partial f}{\partial x_2}, \ldots \frac{\partial f}{\partial x_n}\right)$$

F(x) = Hessian matrix of second derivatives

$$= \begin{pmatrix} \frac{\partial^2 f}{\partial x_1 \partial x_1} & \frac{\partial^2 f}{\partial x_1 \partial x_2} & .. & \frac{\partial^2 f}{\partial x_1 \partial x_n} \\ \frac{\partial^2 f}{\partial x_2 \partial x_1} & \frac{\partial^2 f}{\partial x_2 \partial x_2} & .. & \frac{\partial^2 f}{\partial x_2 \partial x_n} \\ .. & .. & \frac{\partial^2 f}{\partial x_i \partial x_j} & .. \\ \frac{\partial^2 f}{\partial x_n \partial x_1} & \frac{\partial^2 f}{\partial x_n \partial x_2} & .. & \frac{\partial^2 f}{\partial x_n \partial x_{nl}} \end{pmatrix}$$

For a sequence of points $\mathbf{x}_i$ (hopefully converging toward the minimum), denote by g_i, i = 1,2,3..., the sequence of gradient vectors $\mathbf{g}(\mathbf{x}_i)$.

The Nelder-Mead Simplex Algorithm

The simplex method given by Nelder and Mead (1965), sometimes called the *downhill simplex method,* is one of the few robust and efficient methods that does not use any derivative information. This greatly simplifies computational requirements and reduces the chances of errors that can crop up in the differentiation of complex rate expressions.

A *simplex* in n dimensions is the convex body determined by n+1 vertices. Thus, in a 2-dimensional plane a simplex is a triangle determined by its three vertices. In 3D space a simplex is a tetrahedron determined by its 4 vertices, etc. The idea of Nelder and Mead, for an n-dimensional problem, is to start with n+1 points $\mathbf{x}_i$ and geometrically transform and move the simplex they determine until a minimum in $f(\mathbf{x})$ has been reached. The main steps are outlined, in 2-dimensions, in Figure 8.1.

a. The highest point (i.e. where $f(\mathbf{x})$ is highest) is reflected through the opposite face. This new vertex replaces the highest point if it gives good improvement (i.e. has a much lower value of the function $f(\mathbf{x})$)
b. If the reflected point is significantly better that the old highest point, then it is extended further along the same direction as long as this gives additional improvement
c. If the reflected point is only marginally better than the old highest point, then use it. At the same time, contract the simplex by one-half along one dimension toward the opposite face, if this reduces, and hence can replace, the second highest point.
d. If the reflected point is worse than the old highest point, or is only marginally better, and the one-dimensional contraction toward the opposite face does not help, then do an n-dimensional contraction toward the lowest point.

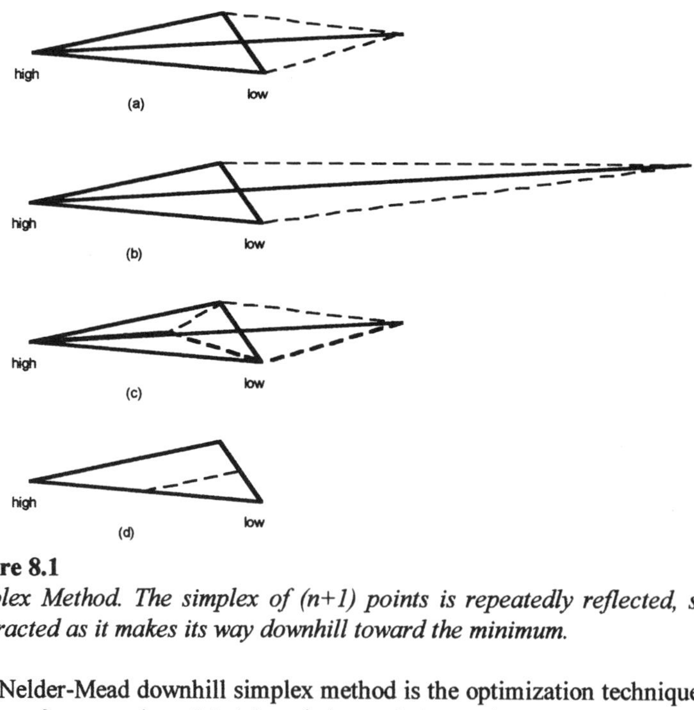

Figure 8.1
Simplex Method. The simplex of (n+1) points is repeatedly reflected, stretched, and contracted as it makes its way downhill toward the minimum.

The Nelder-Mead downhill simplex method is the optimization technique incorporated in the software package Matlab as fmins or fminsearch.

Steepest Descent

Most optimization methods use at least first-derivative information about the objective function. The most natural of these methods, although far from the best, is the method of steepest descent. In this method, one simply starts with any estimate $\mathbf{x}_0$, calculates the gradient $\mathbf{g}(\mathbf{x}_0)$ at that point, and performs a one-dimensional search for the minimum of $f(\mathbf{x})$ in the direction of $-\mathbf{g}(\mathbf{x}_0)$. Since $\mathbf{g}(\mathbf{x}_0)$ points in the direction of most rapid increase of $f(\mathbf{x})$, it follows that $-\mathbf{g}(\mathbf{x}_0)$ points in the direction of most rapid decrease of $f(\mathbf{x})$. The point $\mathbf{x}_1$ where a minimum occurs should be an improvement over $\mathbf{x}_0$. The procedure is repeated starting at $\mathbf{x}_1$, and iterated as often as necessary to achieve the desired accuracy.

This method, while robust, can be very slow. It often requires many iterations to get close to the desired minimum. This can happen even when the objective function is a simple quadratic with two parameters, if its fit contours are highly eccentric and the starting estimate is unfortunately chosen. This is illustrated in two dimensions in Figure 8.2, showing the elliptical contours of equivalent fit for a quadratic objective. The gradient at each point is orthogonal to the contour at that point, and the minimum along that gradient is achieved where that gradient line is tangent to a contour. This results in the zigzag search path shown by the heavy line, with each segment orthogonal to the previous segment. Even in this simple case, with contours that are not very eccentric, it is evident that many iterations are needed to approach the minimum (best fit) at the center of the ellipses.

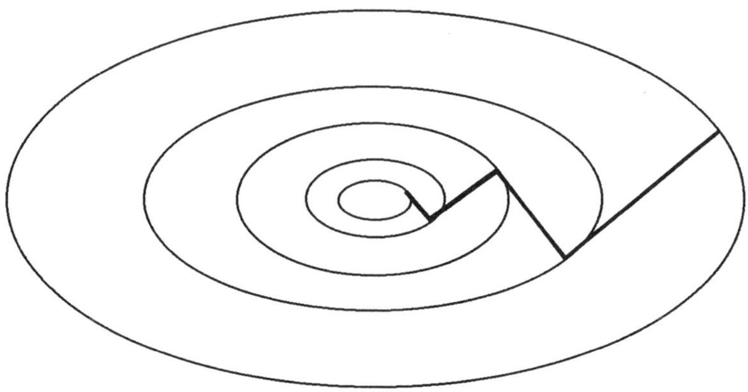

Figure 8.2.
Steepest descent. The successive gradients are at right angles to each other, resulting in a very zigzag path toward the minimum.

Conjugate Gradient Algorithms

In the conjugate gradient method, the idea is again to pick a succession of descent directions $\mathbf{d}_i$ and successively minimize $f(\mathbf{x})$ along them. However, instead of using the gradients $-\mathbf{g}(\mathbf{x}_i)$ as the directions, and obtaining a succession of orthogonal directions, the idea is to construct a succession of so-called "conjugate" directions $\mathbf{d}_i$ that generally lead much faster to the minimum. The procedure is as follows.

When the objective function is a quadratic $f(\mathbf{x}) = \mathbf{x}^t\mathbf{Q}\mathbf{x}$ for some n by n matrix $\mathbf{Q}$, then the directions $\mathbf{d}_i$ for i = 0,1,2,3...are called *conjugate* if they are *Q-orthogonal*, i.e. if $\mathbf{d}_i^t\,\mathbf{Q}\mathbf{d}_j = 0$ for $i \neq j$. It can be analytically shown that, for an n-dimensional quadratic problem, starting with any direction $\mathbf{d}_0$, and using any set of **Q**–orthogonal directions, the minimum will be reached in, at most, n steps. For example, in two dimensions two **Q**–orthogonal directions correspond to conjugate diameters, as shown in Figure 8.3 (where the first direction is taken to be $-\mathbf{g}(\mathbf{x}_0)$).

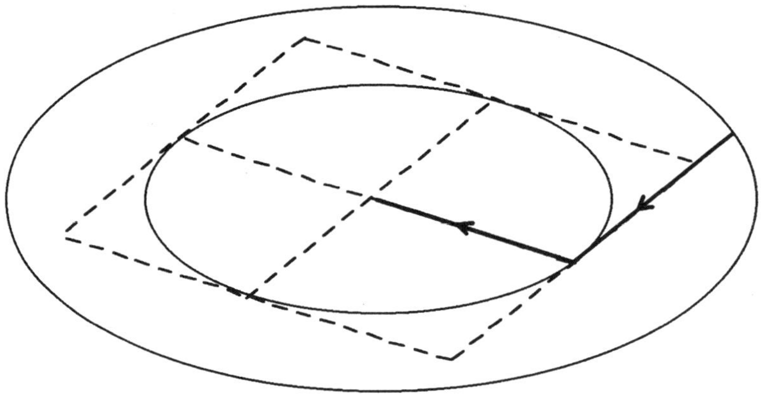

Figure 8.3 Conjugate Directions.
The first direction is the negative gradient, and the second is along the conjugate diameter, leading straight to the minimum.

What makes the method particularly attractive, both in the quadratic case and in the more general non-quadratic case, is that it is not actually necessary to know the matrix **Q** explicitly. A relatively straightforward exercise shows that for a quadratic function the following algorithm produces **Q**-orthogonal directions $\mathbf{d}_k$ and a sequence of points $\mathbf{x}_i$ that converges to the minimum of f.

Start at some $\mathbf{x}_0$.

1. Let $\mathbf{d}_0 \quad = -\mathbf{g}_0 = -\mathbf{g}(\mathbf{x}_0)$ (i.e. start with steepest descent.)
2. Let $\mathbf{x}_{k+1} \quad = \mathbf{x}_k + \alpha_k\mathbf{d}_k$, where α is chosen to minimize $f(\mathbf{x})$ in this direction.
3. Let $\mathbf{d}_{k+1} \quad = -\mathbf{g}_{k+1} + \beta_k\mathbf{d}_k$, where $\beta_k = \mathbf{g}^t_{k+1}\,\mathbf{g}_{k+1} \,/\, \mathbf{g}^t_k\mathbf{g}_k$.
4. Repeat 2 and 3 until convergence has been achieved.

By this means we arrive at a *conjugate gradient* method that can be applied to any function f, that needs only the gradients of f, and is relatively simple to implement. If f is quadratic then convergence will be achieved in n steps at most. If f is not quadratic then the algorithm cannot be expected to converge exactly in n steps. But, if the starting point x_0 is reasonably close to the minimum, then f will be approximately quadratic locally and the algorithm will give rapid convergence toward the minimum.

There is one small manoeuvre that is known to be helpful in the implementation of this method; one could just keep on repeating the steps of finding new points and new directions, etc., but in practice it turns out to be more effective to restart every n steps with a steepest descent step, by taking $\mathbf{d}_n = -\mathbf{g}_n$. The conjugate gradient method describe above is due to Fletcher and Reeves (1964).

A significant modification, that might at first have seemed trivial, was later introduced by Polak and Ribiere (1969) (see Polak (1971)). Using the fact that $\mathbf{g}_k{}^t\mathbf{g}_{k+1} = 0$ for quadratic functions, an equivalent formula for β_k in step 3 above (at least in the quadratic case) is the following:

$$\beta_k = \frac{(\mathbf{g}_{k+1} - \mathbf{g}_k)^t \mathbf{g}_{k+1}}{\mathbf{g}_k^t \mathbf{g}_k} \tag{8.3}$$

Numerical experience indicates that this modification generally produces faster convergence than Fletcher-Reeves for non-quadratic functions as well.

One of the principal advantages of conjugate gradient methods over the Newton-type methods described below (other than their simplicity of implementation) is their ability to handle problems with a large number of parameters, possibly in the thousands. However, this is not much of a consideration in problems of fitting data to kinetic rate expressions, where there are unlikely to be more than a dozen or so parameters to be estimated.

Conjugate gradient is the default optimization method in the software package MathCad. The "Professional" version of MathCad also has other methods built in as options, namely Levenberg-Marquardt and quasi-Newton, both described below. The spreadsheet packages Excel and Quattro-Pro have optimization packages essentially identical to each other (supplied by Frontline Systems, Inc.) that include conjugate gradient as an optional method, although the quasi-Newton method (discussed below) is their default option.

Newton's Method

Newton's method is based on using a truncated Taylor series to approximate $f(\mathbf{x})$ by a quadratic function and then minimizing this quadratic approximation. Thus, near any point $\mathbf{x}_k$

$$f(\mathbf{x}) = f(\mathbf{x}_k) + \nabla f(\mathbf{x}_k)(\mathbf{x} - \mathbf{x}_k) + \tfrac{1}{2}(\mathbf{x} - \mathbf{x}_k)^t \mathbf{F}(\mathbf{x}_k)(\mathbf{x} - \mathbf{x}_k) \tag{8.4}$$

The right side of the above expression is minimized at

$$\mathbf{x}_{k+1} = \mathbf{x}_k - [\mathbf{F}(\mathbf{x}_k)]^{-1} \nabla f(\mathbf{x}_k) \tag{8.5}$$

If the starting point $\mathbf{x}_0$ is reasonably close to the minimum then the sequence $\mathbf{x}_k$ generated by the above recursion will converge very rapidly to the minimum. This very rapid convergence makes Newton's method an attractive method to try to implement. Unfortunately, in its pure form as described above there are several major difficulties.

First, the method assumes the Hessian matrix $\mathbf{F}(\mathbf{x})$ is available. For a problem with n parameters, this involves correctly calculating and coding approximately $n^2/2$ second derivatives. For an n of even moderate size it seems to be impossible to do this successfully, even with the help of modern symbolic software. One could of course use numerical differencing to estimate the necessary second derivatives, but then the amount of computation required exceeds that of other more radical revisions of Newton's method, such as the quasi-Newton methods described below.

Another difficulty when n is large is the storage and computational burden of calculating $\mathbf{F}(\mathbf{x})$ and $[\mathbf{F}(\mathbf{x})]^{-1}$. This is less of a issue than it used to be and remains a difficulty only for problems with many hundreds of parameters. Fortunately, no problem involving fitting TSR data to rate expressions has anything like this number of parameters.

A more serious difficulty with the pure Newton's method is that the recursion given above may not converge at all if the starting point $\mathbf{x}_0$ is not close to the minimum. To help solve this problem, the vector $\mathbf{d}_k = -[\mathbf{F}(\mathbf{x})]^{-1}\nabla f(\mathbf{x}_k)$ is not used simply as a correction vector, but rather as a descent direction along which f is minimized. Even this modification does not suffice in itself, since there is no guarantee that $\mathbf{d}_k$ will be a descent direction when $\mathbf{x}_k$ is far from the minimum. The rapid convergence one hopes for depends on $\mathbf{F}(\mathbf{x})$ being positive definite; this will be the case near a minimum but need not be so further away. In practice, then, some major modifications of Newton's method are needed to achieve a successful fitting routine. Two of these are described in the next two sections.

Levenberg-Marquardt Method

One possible modification of pure Newton's method is to replace the search direction $-[\mathbf{F}(\mathbf{x}_k)]^{-1}\nabla f(\mathbf{x}_k)$ by $-[\varepsilon_k\mathbf{I}+\mathbf{F}(\mathbf{x}_k)]^{-1}\nabla f(\mathbf{x}_k)$, where ε_k is a positive parameter that produces a compromise between steepest descent (when ε_k is very large), and Newton's original method (when ε_k is zero). The aim is to achieve the robustness of steepest descent (applicable when one starts far from the minimum) by an appropriately large choice of ε_k, and subsequently achieve the convergence speed of Newton's method (once one gets close to the minimum) by choosing a smaller value of ε_k. Note that it is always necessary to choose ε_k so that $[\varepsilon_k\mathbf{I} + \mathbf{F}(\mathbf{x}_k)]^{-1}$ is positive definite.

There are various algorithms for finding a suitable ε_k. These vary in difficulty. One method widely used because of its relative simplicity is due to Marquardt (1963) based on ideas first proposed by Levenberg (1944). The idea is to start with some value of ε_k and attempt to factor $\varepsilon_k\,\mathbf{I} + \mathbf{F}(\mathbf{x}_k)$ as a product $\mathbf{G}\mathbf{G}^t$ for some lower triangular matrix $\mathbf{G}$ having strictly positive diagonal elements. This is the well-known Cholesky factorization [c.f. Luenberger, 1983, p. 194], which is possible if, and only if, $\varepsilon_k\,\mathbf{I} + \mathbf{F}(\mathbf{x}_k)$ is positive definite, and which is easily carried out when the factorization is possible. If the factorization is not possible, then ε_k is increased until it is. It is then easy to calculate $\mathbf{d}_k = -[\varepsilon_k\mathbf{I} + \mathbf{F}(\mathbf{x}_k)]^{-1}\,\nabla f(\mathbf{x}_k)$ by successively solving the two simple triangular systems:

$$\begin{aligned} \mathbf{G}\mathbf{y} &= -\mathbf{g}_k \\ \mathbf{G}'\mathbf{d}_k &= \mathbf{y} \end{aligned} \qquad (8.6)$$

After this is done, the values of $f(\mathbf{x}_k + \alpha\mathbf{d}_k)$ are examined, and on this basis $\mathbf{x}_{k+1}$ is determined and ε is either increased or decreased.

The Levenberg-Marquardt method is often recommended for least-squares problems since the special structure of these problems allows for a simple approximation to the Hessian matrix using only first derivatives (Fletcher (1987), pp110-112). The method is available as one of the minimization options in MathCad and often found in other packages.

Quasi-Newton Methods

Quasi-Newton methods attempt to achieve the very fast convergence of Newton's method without having to calculate the Hessian matrix explicitly. The idea is to use gradients to successively build up an approximation to the inverse Hessian. For Newton's method, new directions are taken as

$$\mathbf{d}_k = -[\mathbf{F}(\mathbf{x}_k)]^{-1}\nabla f(\mathbf{x}_k) \tag{8.7}$$

so, more generally, we might take new directions as

$$\mathbf{d}_k = -\mathbf{H}_k \nabla f(\mathbf{x}_k) \tag{8.8}$$

for some well-chosen symmetric matrix $\mathbf{H}_k$. It is easy to show [see Luenberger (1984) pp 261-266 for this and other details of the quasi-Newton methods] that there will be good convergence when $\mathbf{H}_k \cong \mathbf{F}^{-1}$. For convenience denote

$$\begin{aligned} \mathbf{p}_k &= \mathbf{x}_{k+1} - \mathbf{x}_k \\ \mathbf{q}_k &= \mathbf{g}_{k+1} - \mathbf{g}_k \end{aligned} \tag{8.9}$$

Then it can be shown that $\mathbf{F}\mathbf{p}_k \cong \mathbf{q}_k$, so $\mathbf{F}^{-1}\,\mathbf{q}_k \cong \mathbf{p}_k$. Thus we want $\mathbf{H}_k\mathbf{q}_k \cong \mathbf{p}_k$. The basic procedure is to start with any symmetric positive definite $\mathbf{H}_0$, say, $\mathbf{H}_0 = \mathbf{I}$, and try to update $\mathbf{H}_k$ to $\mathbf{H}_{k+1}$ in such a way that

$$\mathbf{H}_{k+1}\,\mathbf{q}_k = \mathbf{p}_k \tag{8.10}$$

There are many different ways to update $\mathbf{H}_k$ to satisfy this basic recursion requirement. The simplest is the rank one correction, described below.

Rank one correction

Try the form $\mathbf{H}_{k+1} = \mathbf{H}_k + \alpha - \mathbf{u}\mathbf{u}^t$, noting that $\mathbf{u}\mathbf{u}^t$ is a symmetric rank-one matrix. It turns out that the basic recursion requirement is satisfied by taking $\mathbf{u} = \mathbf{p}_k - \mathbf{H}_k\mathbf{q}_k$, and

$$\alpha = \frac{1}{\mathbf{q}_k^t(\mathbf{p}_k - \mathbf{H}_k\mathbf{q}_k)} \tag{8.11}$$

This method can be used successfully, but has some serious difficulties. There is no guarantee that $\mathbf{H}_{k+1}$ will remain positive definite (because the value of α might be negative). Also, the denominator in the formula for α can easily be very small, causing severe numerical difficulties. To avoid these difficulties one may try rank-two corrections, described below.

Davidon-Fletcher-Powell (DFP) method

The first generally successful method was presented by Davidon (1959), who called it a *variable metric* method. It was subsequently developed by Fletcher and Powell (1963). They show, and it is straightforward to verify directly, that the following rank two correction satisfies the basic recursion requirement.

$$\mathbf{H}_{k+1} = \mathbf{H} + \frac{\mathbf{p}\mathbf{p}^t}{\mathbf{p}^t\mathbf{q}} - \frac{\mathbf{H}\mathbf{q}\mathbf{q}^t\mathbf{H}}{\mathbf{q}^t\mathbf{H}\mathbf{q}} \tag{8.12}$$

where the subscript k on all the right hand terms have been dropped for convenience.

This rank two correction has no numerical problems with small denominators, and it can be shown that $\mathbf{H}_{k+1}$ is always positive definite if $\mathbf{H}_k$ is. This guarantees that $\mathbf{d}_k$ will always be a descent direction, thus overcoming one of the serious difficulties of the pure Newton method. The Davidon-Fletcher-Powell method works quite well, but it turns out that the slight modification below gives experimentally better results, even though it is theoretically equivalent.

Broyden-Fletcher-Goldfarb-Shanno (BFGS) method

The BFGS correction formula was discovered independently and more-or-less simultaneously by Broyden (1970), Fletcher (1970), Goldfarb (1970) and Shanno (1970). The idea is to pretend one is using a DFP-type scheme to estimate $\mathbf{F}$ instead of $\mathbf{F}^{-1}$, i.e. take $\mathbf{F}_0 = \mathbf{I}$ and try to get $\mathbf{F}_k\,\mathbf{p}_k \cong \mathbf{q}_k$. The update scheme would be the same as above with **p**'s and **q**'s interchanged. Again, the subscript k is omitted from every term on the right side, and we write:

$$\mathbf{F}_{k+1} = \mathbf{F} + \frac{\mathbf{q}\mathbf{q}^t}{\mathbf{q}^t\mathbf{p}} - \frac{\mathbf{F}\mathbf{p}\mathbf{p}^t\mathbf{F}}{\mathbf{p}^t\mathbf{F}\mathbf{p}} \tag{8.13}$$

Of course, what is really required is $\mathbf{H}_{k+1} = \mathbf{F}^{-1}{}_{k+1}$. But, assuming $\mathbf{H}_k = \mathbf{F}_k^{-1}$ has already been obtained, one can use a standard (but messy) formula for inverses of rank two perturbations to invert the expression for $\mathbf{F}_{k+1}$. Happily in this case most of the mess cancels out, and one arrives at the BFGS update (where again the subscripts k on the right side have been omitted):

$$\mathbf{H}_{k+1} = \mathbf{H} + \left(1 + \frac{\mathbf{q}^t\mathbf{H}\mathbf{q}}{\mathbf{p}^t\mathbf{q}}\right)\frac{\mathbf{p}\mathbf{p}^t}{\mathbf{p}^t\mathbf{q}} - \left(\frac{\mathbf{p}\mathbf{q}^t\mathbf{H} + \mathbf{H}\mathbf{q}\mathbf{p}^t}{\mathbf{p}^t\mathbf{q}}\right) \tag{8.14}$$

This BFGS update is only slightly more complicated than the DFP update, and this very slight extra computation is more than compensated for by the extra speed of convergence that is observed in practice.

As mentioned previously, quasi-Newton is the default non-linear optimization method included in Excel and Quattro-Pro, and is an option in the Professional Version of MathCad and in the Optimization Toolbox for Matlab. Various correction methods are made available in different packages.

Summary of Optimization Methods

The Nelder-Mead downhill simplex algorithm has the advantage of being very reliable and requiring no derivative evaluations. On the other hand, it is slower than methods that do use derivative information. Of course, even a relatively slow method may be quite fast enough for a not-too-large problem on a fairly fast computer.

Steepest descent by itself is far too slow to be a practical method in view of the complexities that are often encountered in modern data fitting, although it has an honourable history, having been used by Gauss himself. However, it does guarantee convergence to a minimum, and so it is often used for starting steps and periodic restarting steps in faster algorithms.

Conjugate gradient methods are fast, though not as fast as Newton-type methods near the minimum. They can handle large numbers of parameters since their storage requirement (needing only to store gradients) is a small fraction of that for Newton-type methods (which need to store Hessian matrices or their approximations). These methods are also relatively simple to implement, but this is of little concern to most practitioners who are unlikely to write their own optimization code.

Newton-type methods are very fast near the minimum. However, pure Newton methods can have serious convergence difficulties if the starting point is far from the minimum. Hence some modified form is nearly always required in practice. For least squares problems a Levenberg-Marquardt method is attractive, since it guarantees convergence by balancing steepest descent with Newton's method, and since the special structure of least squares problems allows for an easy approximation to the Hessian matrix.

Quasi-Newton methods, usually using the BFGS update, are the methods of choice for non-linear optimization, especially for problems with a small to medium number of parameters to be estimated, as is the case with kinetic rate data fitting. They converge with close to the speed of pure Newton methods, without the convergence uncertainties of Newton's method, or the need to compute second derivatives. Their main drawback of requiring storage and inversion of possibly very large matrices only applies when a very large number of parameters (many hundreds) have to be estimated.

The recommended methods for fitting TSR data are therefore two:

- the Nelder-Mead simplex algorithm
- the Quasi-Newton method using the BFGS update.

The choice will depend on the complexity of the rate equation, the speed of convergence and the capabilities of the available computers. Unfortunately even the best of available optimization methods does not guarantee success in optimizing fits of kinetic rate expressions to plentiful data. The form of many kinetic rate expressions, particularly those encountered in catalysis, guarantees that difficulties will arise and brings out the need for a careful, thoughtful approach to data fitting.

Choosing a Data Fitting Procedure

Choosing a suitable minimization procedure, be it conjugate gradient, quasi-Newton or some other, does not in itself guarantee success in identifying a unique rate expression or getting good parameter estimates when fitting data to the chosen rate expression. Because of the high correlation among the various parameters in rate expressions (as discussed in Chapter 4), a poor choice of starting values can easily lead to a local minimum of the sum of squares of residuals. Such local minima can lie far from the global minimum and the true parameter values.

A usually reliable procedure for finding good initial values (starting estimates, see Watts, 1994) is to begin by fitting the rate expression for various fixed temperatures T_j, i.e. begin with a series of traditional isothermal fittings. For each fixed temperature T_i the rate constants $k_j = A_j \exp(-E_j / RT_i)$ are then obtained as simple constants k_i, not involving the touchy exponential dependence of the Arrhenius plot. Then, for each k_i, the fitted values $k_i(T_j)$, and the corresponding temperatures T_j, can be used to form an Arrhenius plot. This will yield a fair set of initial estimates of the frequency factors A_i and activation energies E_i for each of the constants of the rate expression. This presumes that the Arrhenius relationship applies to all fundamental kinetic and thermodynamic constants that appear in rate expressions, which it does to a good approximation, if the constants that appear in the rate expression are indeed fundamental.

These estimates of A_i and E_i will not themselves be good enough to be the final estimates we will accept in TSR work, but they are much better than randomly chosen starting estimates for a subsequent general non-linear least-squares fitting procedure of the whole data set. This procedure fits the candidate rate expression, with the Arrhenius parameters supplied by the isothermal fits, to the same rate expression written to include the temperature dependence of all the constants. The overall fit is then optimized using temperature centering (see Chapter 4) and the complete range of temperatures and conversions studied. The results are then examined to see if a satisfactory fit has been obtained. Chapter 11 illustrates some fits.

In fitting data one should be aware of the merits of the different data treatment methods, among them different optimization methods. When necessary, one should apply a series of different optimization methods, varying step sizes, starting values and other available parameters to search the parameter space for global and local minima. At the end of such a search our understanding of physical realities, extraneous information and chemical intuition, should be brought to bear.

> *In pursuing a theoretical understanding of a reaction it is not necessarily wrong to select a less-than-best fit as the one meriting further attention if other considerations so indicate.*

It is quite possible that the fit with the least sum of squares is not the fit that merits acceptance. It is possible to imagine a fit whose SS is smallest but whose errors are not uniformly distributed over the range of conditions investigated. It is generally true that this feature of error distribution should be, but often is not, considered and used to arrive at a final decision regarding the selected rate expression or parameters.

This introduces a complication for the interpreter. How to justify the choice of a less than minimal-fit set of parameters? Such a choice might be defended by evidence from more extensive studies, preferably done by someone else. A more satisfactory solution would be to find evidence from other-than-kinetic experiments. Such auxiliary information may come from theoretical calculations, adsorption studies or some other source, as long as its pertinence to the kinetic parameters is well understood. In using the data for reactor design, however, the best fit is normally the preferred choice for further use.

This difficulty once again brings into play the question of judgment on the part of the interpreter. It appears that complete objectivity in data interpretation is elusive, or even impossible (see Bates and Watts, 1988, pp. 86-131). The methods described above and in Chapter 7 are useful, even essential, tools in the process of arriving at an interpretation of experimental observations.

> *In the end, however, the intuition of the interpreter is still needed to formulate the initial hypothesis that will start the process of data interpretation.*

This will be subject to extensive testing as more data accumulates, other workers try to reproduce and extend the results, and ancillary information is brought to bear on the issue. During this process it is important not to let unverified or scant contrary results dominate discussion on the validity of a hypothesis. Contrary results can be indicative of several problems. For example:

- one of the studies is in error;
- the two studies are not dealing with comparable systems;
- the current understanding is not broad enough and needs to be altered to accommodate both studies.

Until such issues have been resolved, one need not reject an established understanding. On the other hand, they must not be simply ignored; a resolution is essential. It is bad science to ignore contrary evidence, fail to cite it, and fail to deal with it adequately.

9. Interpretation of Rate Parameters

Numbers by themselves have no physical meaning. Numbers describing physical quantities do have meaning but only within certain constraints. Numbers assigned to such physical quantities have meaning and are valid only within these constraints.

The Parameters Involved in Rate Expressions.

The five *fundamental* kinetic parameters of rate expressions are:

- the *activation energy*, ΔE, appearing in the exponent of the Arrhenius expression;
- the *frequency factor* or *pre-exponential*, A, appearing before the exponential term of the Arrhenius expression;
- the *enthalpy* of adsorption, ΔH, which appears in the exponent of adsorption equilibrium constants in catalytic rate expressions;
- the *pre-exponential*, which in adsorption equilibrium constants is usually written as an entropy of adsorption, ΔS in exponential form;
- the *reaction order* with respect to each of the reactant concentrations appearing in the rate expression.

Quantities corresponding to the above are frequently calculated from an Arrhenius plot for the parameters of *empirical* rate expressions. The sizes of these are not constrained by theory. The size of all *fundamental* kinetic parameters, in contrast, is invariably constrained by physical realities. On the low side the limit is usually zero. The high side is constrained, if only by the simple impossibility of accepting infinite values for approachable physical quantities. The reasons for the existence of limits less extreme but no less definite than zero and infinity must be understood, so that parameters which fall beyond realistic limits can be spotted in fitted rate expressions. Parameters that are out of realistic range cast serious doubt on the validity of the mechanism behind the rate expression being fitted.

The Fundamental Constraints on Activation Energies

Unimolecular Elementary Reactions

The energy required for the breaking of a chemical bond in a reaction cannot be lower than, nor significantly higher than, the strength of the bond itself. This seems self evident. In *unimolecular* reactions at least enough energy to cause the desired conversion of the molecule must be in the molecule as the reactant passes through the *transition state* between reactant and product. The transition state is a condition in which the reacting molecule not only contains enough energy to break (or otherwise alter) the bond in question, but the energy is focused on the bond in question so that it can force the molecule to turn from reactant to product. In a *scission* reaction the bond in question achieves zero strength at this point. Clearly then, the minimum energy required for the scission is that necessary to break the bond. One might suspect that the total energy that the molecule contains may not be efficiently focused on the breaking bond, so the energy the molecule requires for this event to take place may have to be somewhat higher, on average, than the absolute minimum required for bond breakage. However, theoretical considerations lead to the conclusion that, given enough time, a minimal energization will come to a focus on the weakest bond and break it. In other words, the time between energizing/de-energizing collisions is found to be sufficiently long to make this event possible with not much more than the minimum energy, and the approximation that the activation energy for bond breakage is near the corresponding bond strength is usually valid. Table 9.1 gives some idea of the bond strengths of organic molecules.

All *elementary* reactions resulting in bond breakage are therefore limited on the low side in activation energy by the strength of the bond that will be broken. This is generally also the weakest bond in the molecule. Table 9.1 gives some idea of what the required activation energy might be for elementary reactions breaking the bond indicated. To get to this level of energy content, the molecule must rise high above the energy content of an average molecule in the reacting mixture.

In many cases, however, the bond of interest is not the weakest bond in the molecule and the reaction breaking it can never occur; the weakest bond will break instead. This is an immutable fact in the case of elementary reactions, but nature often provides a *mechanism* that allows the overall conversion to take place at a lower cost of energy and, in the case of catalysis for example, at specific bonds, regardless of bond strength.

Table 9.1
Bond Strengths of Selected Organic Molecules

Bond	Strength in kJ/mol
$H\text{-}CH_3$	430
$H\text{-}C_6H_5$	460
$H\text{-}CN$	500
$H\text{-}NH_2$	460
$H\text{-}OCH_3$	430
$CH_3\text{-}CH_3$	370
$CH_3\text{-}C_6H_5$	420
$C_6H_5CH_2\text{-}CH_3$	300
$C_6H_5CH_2\text{-}OH$	330

Monomolecular reaction mechanisms

A reaction mechanism consists of a sequence of events that constitute an integrated path from reactants to products. For example, we can have a homogeneous reaction mechanism or a catalytic reaction mechanism that will allow a lower energy path to products than the one available via an elementary bond breaking reaction. Chain mechanisms in homogeneous reactions, and on some catalysts, make possible processes that allow strong bonds to be broken or rearranged by means of pathways that are initiated by a full-energy-cost breakage of a bond, and proceed to break or rearrange bonds in subsequent molecules via a complex series of events but at a lower energy cost. Both homogeneous and catalytic mechanisms operate by means of intermediate propagating reactions that lower the energy cost by introducing sequential events in the several propagation steps of the mechanism.

An example of a monomolecular reaction mechanism is presented by the pyrolysis of ethane to form ethylene and hydrogen. The molecular elimination of hydrogen from ethane does not take place in a single elementary reaction, for reasons involving entropy considerations that we need not discuss here. Suffice it to say that the direct production of hydrogen and ethylene has not been observed under the conditions involved in pyrolysis. What happens instead, as a molecule is energized by collisions, is that the weakest bond in the molecule breaks. The weakest bond in this case is the C–C bond. The cracking of the C–C bond therefore constitutes the *initiation* step and this requires as much energy as it takes to break the C–C bond in ethane.

This event is followed by a series of propagating steps, with the overall result that ethylene and hydrogen are produced. The mechanism is such that many of these low-energy conversion events (*propagation* reactions) take place per initiating event, distributing the energy cost of the initiation event among the numerous consequent conversion events that take place in the propagation steps. The overall activation energy of such a means of breaking ethane into ethylene and hydrogen ends up being between the low energies required in the propagation steps and that of the strength of the bond broken in the initiating step. Chapter 3 contains a more detailed description of this process.

Exothermic monomolecular reactions

Exothermic reactions can, in principle, have as little as zero activation energy, although compounds with such a property would be very unstable. As a result, even exothermic reactions require some activation energy to get going. However, following initiation, exothermic reactions can be explosive, since the energy released during reaction raises the temperature and hence the rate of reaction, and hence the rate of heat release and so the temperature, and so the rate of reaction, and so on.

Guidelines for activation energies of monomolecular reactions

The upshot is that activation energies for overall conversion are expected to fall between ΔE_{max} of about 100 and ΔE_{min} of 0 kcal/mol (~400 and 0 kJ/mol) regardless of the mechanism of reaction. If the bond strengths in a molecule are known, the expected range of activation energies in a homogeneous non-catalytic reaction can often be narrowed to a span of 10 kcal/mol (~40 kJ/mol) within that range. Moreover, taking into account the instability of low-activation energy reactions and the slowness of high-activation energy reactions, most processes encountered in practice have activation energies lying between 30 and 70 kcal/mol (120 to 280 kJ/mol).

In catalytic reactions the range of activation energies is lowered by the fact that the initiating intermediate forms a weakly-bonded chemisorbed species whose attachment to the surface is rarely beyond 50 kcal/mol (200 kJ/mol). This limits the expected catalytic activation energies to between 20 and 50 kcal/mol (80 to 200 kJ/mol) and makes catalytic reactions less temperature sensitive than non-catalytic processes. Moreover, catalytic reactions can have such low activation energies that a pinch of catalyst may set off an explosion in an exothermic reaction, even if the reaction will not proceed at all in the absence of the catalyst.

A most important point in all of this is that *the activation energy cannot be negative.* Any data-fitting that results in such a parameter in a rate expression is unacceptable for interpreting the mechanism of the reaction, for the above-discussed reasons. Even if such a fit is useful for practical purposes, there is no sense in offering a theoretical interpretation of a negative value of the activation energy.

Bimolecular Elementary Reactions

Similar arguments hold in the case of bimolecular elementary reactions. The activation energy required for reaction is supplied mainly by the internal energy of the reacting molecules. Only in elementary reactions such as those of simple diatomic molecules is it the collision energy alone that supplies the activation energy of bimolecular reactions. The reaction itself once again involves bond breakage as the energy-demanding step in the process, although in bimolecular reactions some of this energy may be compensated by simultaneous bond formation at another site. In fact, bimolecular reactions are often exothermic, so that the activation energies in elementary bimolecular reactions will, on average, be lower than those for unimolecular processes. All in all, it is safe to say that the upper limit of activation energy in bimolecular reactions is no higher than the limit for unimolecular reactions discussed above.

It is difficult, if not impossible, to estimate the activation energy of elementary bimolecular reactions from bond strength data, or any other readily available source. With increasing frequency it can be estimated by various methods of fundamental calculation of molecular bonding, but even then the intra-molecular mechanism of the reaction is not always clear and has a significant effect on the activation energy being calculated. Most calculated activation energies are meant to check the measured quantities rather than attempting to predict them.

Fundamental Constraints on Frequency Factors

Unimolecular Elementary Reactions

The *frequency factor*, or pre-exponential, of a rate constant represents the theoretical rate at which the reaction would proceed at unit concentration of reactants when the activation energy requirement is zero. This is also the rate at which the reaction would proceed at infinite temperature where the exponential term of the Arrhenius equation tends to a value of one. In either case it represents a realistic upper limit of the reaction rate. One can interpret the frequency factor as an upper limit on the speed with which events take place in a reaction. The reason for this limit is that the reaction cannot proceed any faster than the necessary physical rearrangements can take place in a molecule.

Molecules consist of finite masses that vibrate within certain limits as if connected by perfect springs. A reaction event such as the breaking of a bond can occur at most only once per vibration, when the extension of the spring exceeds its breaking strength. This vibrational frequency has been measured to be about 10^{13} sec^{-1} for bonds in organic molecules, a value that therefore represents the upper limit of bond breaking frequency. This value has been accepted as the *universal frequency factor* for unimolecular elementary reactions and represents the upper limit of unimolecular frequency factors.

Since most elementary unimolecular reactions require that the molecule must be in an appropriate configuration while breaking a bond, the actual frequency factor is usually lower than the universal value. The factor reducing the upper limit is by agreement thought of as the *stearic factor* and represents an entropy requirement for achieving the transition state. When the stearic factor is expressed, not by a simple number (P), but by an entropy of activation (ΔS) it can be written as:

$$A = 10^{13}P = 10^{13}\exp(\Delta S/R)\ sec^{-1}. \tag{9.1}$$

Alternatively the whole frequency factor can be written in terms of entropy:

$$A = 10^{13}P = \exp(\Delta S'/R)\ sec^{-1} \tag{9.2}$$

The two ways of defining the entropy of activation can be confusing; we will concentrate on the first. The entropy term then represents a constraint that prevents the reaction from achieving its maximum rate. In this view, the greater the *negative* entropy of activation, the more organized must be the transition state as reactants become products. It is easy to grasp this concept; the greater the degree of organization required for a molecule to make the transition from reactants to products, the lower is the frequency with which it will achieve the necessary configuration and make this transition. It is rare that a specific highly-organized transition state is achieved by random vibrations, even in the case of a fully energized molecule.

The second definition requires a *positive* entropy, $\Delta S'$, of activation. In this view the more organized the transition state the less positive entropy is required for reaction. The universal frequency factor therefore represents maximum disorganization in the transition state. The rate is maximized in this view when disorder is maximized.

These are two opposite interpretations of the entropy effects. The second definition requires a limit to the size of the entropy in order not to exceed the universal frequency factor. All in all, it is the first interpretation that is the most informative.

Unimolecular frequency factors, unlike activation energies, cannot be estimated from other data. There is some chance that one can estimate frequency factors by comparison to the frequency factors of similar reactions.

Bimolecular Elementary Reactions

In bimolecular elementary reactions the frequency factor is limited by the fact that the reacting molecules must collide for reaction to take place. The maximum rate of reaction cannot therefore be greater than the frequency of collisions between molecules. This can be calculated from the kinetic theory of gases to be no higher than about 10^{14} cc/mol/sec for the collision of two atoms at room temperature, and less for the collisions of complex molecules. Collision frequencies are weakly dependent on temperature and can be taken to be constant for all practical purposes. Thus the maximum frequency factor can be taken as 10^{14} cc/mol/sec.

Once again, any decrease from the maximum collision frequency can be ascribed to a negative entropy of activation, this time stemming from the requirement that complex molecules must collide in a specific orientation for a given reaction to take place. Bimolecular frequency factors cannot be estimated from other data except by comparison to the frequency factors of similar reactions.

Frequency Factors and Activation Energies in Mechanisms

Activation Energy

The remaining issue in the formulation of Arrhenius parameters concerns the size of combined activation energies and frequency factors that often appear in overall rate constants of mechanistic rate expressions. Since mechanistic constants are much more commonly encountered than elementary constants, we need to know if they too are subject to limits such as the ones described above. For example, in the pyrolysis of ethane (see Chapter 3) the mechanistic rate expression is:

$$-r_{C2\,H6} = (k_1k_3k_4/k_5)^{1/2}\, C_{C2\,H6} \tag{9.3}$$

Although the overall process is first order and could lead one to believe that the reaction proceeds via a simple unimolecular elimination of hydrogen from ethane, the mechanism is in fact more complex. Conversion proceeds via a chain mechanism involving several steps. The experimental first order rate constant of the overall reaction is composed of the various elementary rate constants, corresponding to several elementary reactions in the overall mechanism, and is subject to the restrictions described above. In such cases one might suspect that the overall activation energies will also be similar to those of elementary rate expressions. This is so in terms of magnitudes.

An examination of the overall rate constant in equation 9.3 shows that the activation energy can be no higher than $(\frac{1}{2})(3)\Delta E_{max} = 1.5\Delta E_{max}$, where ΔE_{max} is the minimum bond strength in the molecule. However, even this would be a very unusual situation, requiring that each of the constants in the numerator have an activation energy of ΔE_{max} and that ΔE_5=0.

In general what tends to happen is that overall constants containing several elementary constants have normal or even low activation energies and frequency factors. The reason is that the elementary propagation reactions of mechanisms, whose rate constants appear in the overall experimental rate constant (k_3 and k_3 here), tend to have low activation energies. In the pyrolysis of ethane the overall activation energy is in fact about 305 kJ/mol, significantly lower than the 370 kJ/mol strength of the C–C bond broken in the initiation step. It is generally true in chain mechanisms that the overall activation energy is lower than the activation energy of the initiating step, guaranteeing that overall activation energies will conform to the limits developed for elementary reactions.

As in all rate expressions, there are no known cases where a mechanistic experimental activation energy is negative. Although the observation of a *negative temperature coefficient* for the rate of the overall reaction is possible (see Chapter 11), the observation of a negative activation energy for a mechanistic rate constant is not.

Frequency Factors

Similar arguments lead to similar conclusions regarding the overall frequency factor, although not as strongly. Somewhat higher than expected frequency factors do occur because it is possible to have unusually low frequency factors in the denominator. In ethane pyrolysis the overall experimental frequency factor is about 10^{15} sec^{-1}. Since frequency factors vary greatly between reactions, one might expect experimental frequency factors to vary from the limits described above by a factor of 10^3 or so. Nevertheless, serious departures from the limits defined above for elementary reactions should as a rule be treated with suspicion if they are found in complex rate constants.

> *In the case of both frequency factors and activation energies, the rate expression normally yields aggregate values for these parameters that represent an averaging of corresponding values of the fundamental rate constants included in the mechanistic rate expression. Such averages are therefore expected to fall within the limits of the extreme values of the component fundamental rate constants.*

Fundamental Constraints on Heat of Adsorption

Adsorption phenomena are involved in the rates of catalytic reactions. They govern the formation of surface intermediates as well as the rates of surface coverage in the standalone processes of adsorption/desorption. Although it is very likely that most catalyst surfaces have adsorption sites with a distribution of adsorption site strengths, the Langmuirian description of surface coverage is almost universally appropriate in describing the rates of catalytic reactions. This formulation states that the fraction of the available sites covered at a given gas phase concentration of adsorbent and at a given temperature is given by the following isotherm equation:

$$\theta = K_A C_A/(1 + K_A C_A) \qquad (9.4)$$

where C_A is the concentration of gas phase adsorbent A;
K_A is the equilibrium constant for adsorption of the species A.

Elaborations of this relationship do not change the fact that some kind of adsorption constant appears in the numerator and denominator of a catalytic rate expression. In these rate expressions the equilibrium constants describe the attachment of reacting molecules, and those of other species in the reacting mixture, to sites on the catalyst surface where reaction will take place: the *active sites*. The rate constant in these cases represents chemisorption, rather than physisorption, and obeys the thermodynamic formulation of equilibrium constants:

$$K_{eq} = k_{ads}/k_{des} = \exp(-\Delta G_a/RT) = \exp(-\Delta H_a/RT)\exp(\Delta S_a/R) \quad (9.5)$$

where ΔG_a is the Gibbs free energy of adsorption;
ΔH_a is the enthalpy of adsorption;
ΔS_a is the entropy of adsorption;
k_{ads} is the rate constant for adsorption;
k_{des} is the rate constant for desorption.

Since most *activated* adsorption on catalysts involves chemisorption, the heat of adsorption, ΔH_a, is generally negative, so that the exponential term containing $-\Delta H_a$ becomes positive and decreases as temperature increases. The physical result is that the surface is more sparsely covered at higher temperatures. Here then is an example of a negative temperature coefficient for the overall coverage, but the activation energies of the two rate constants for adsorption and desorption remain invariably positive. The overall negative temperature coefficient for exothermic adsorption stems from the fact that $|\Delta E_{des}|>|\Delta E_{ads}|$.

In terms of the quantity of heat released upon chemisorption, the amount cannot be higher than the heat released by normal chemical bond formation. In practice it is usually less than that of homogeneous exothermic chemical reactions, say between 10 and 50 kcal/mol (~40 to 200 kJ/mol).

Fundamental Constraints on the Entropy of Adsorption

When we come to consider the entropy term in the adsorption equilibrium constant, there are no well established limits such as those that were present by virtue of vibrational and collision frequencies in the case of elementary rate constants. There are, however, some helpful guidelines. We can begin by examining the ratio k_{ads}/k_{des} as a ratio of two elementary rate constants, both subject to physical constraints of the kind discussed above. The adsorption rate constant frequency factor cannot be any higher than the frequency of collisions between gas phase molecules and active sites.

This number can be calculated from the kinetic theory of gases and turns out to be somewhat lower than the collision frequency between molecules in the gas phase, but we can take it to be the same, about 10^{14} cc mol^{-1} sec^{-1} at room temperature and 1 bar. This high value is usually much reduced in adsorption by site accessibility and "sticking" probability. These concepts account for the fraction of collisions that can access the adsorption site and, on collision, are correctly oriented to allow adsorption to occur.

The frequency factor for desorption is that of a unimolecular decomposition and so is bounded by the frequency of vibration of the bond between the adsorbent and the adsorbate, 10^{13} sec^{-1}, reduced by the usual entropy requirements. This line of reasoning unfortunately gives no clue as to whether the entropy of adsorption or that for desorption is greater, and therefore does not lead to useful guidelines for the *equilibrium* entropy of adsorption.

A second approach attempts to relate the entropy of adsorption to the loss of mobility when a gas phase species, with three degrees of freedom, forms a two dimensional fluid on the surface. This line of reasoning leads to the conclusion that the entropy of adsorption must be negative and no larger than the total entropy of the adsorbate in the gas phase. The authors of this idea (M. Boudart et al. (1967)) go so far as to propose that

$$12.2 - 0.0014\Delta H_a > -\Delta S_a > 10 \text{ e.u.}$$

is a reasonable set of limits on the entropy of adsorption.

Unfortunately the data supporting this conjecture is sparse. The most that can be said is that there indeed seems to be a compensation effect operating between the entropy and enthalpy of adsorption. This leads to increased negative entropy of adsorption as the enthalpy of adsorption increases. The observation raises the question of the contribution to entropy changes of the configurational changes that might take place on adsorption. Such changes are over and above those due to the simple loss of degrees of freedom inherent in condensation from a three-dimensional to a two-dimensional fluid.

Perhaps the only recourse is to search the very sparse data on chemisorption equilibrium constants available in the literature. Figure 9.1 shows a collection of such values, taken largely from Figure 1 in the above Boudart reference and augmented with some additional data. The collected data is plotted on a *compensation effect* plot. This presentation clearly shows that a recognizable compensation effect is operating in activated chemisorption. The evidence points therefore to the supposition that, the higher the heat release on adsorption, the more disorderly will be the adsorbed species. This is an intuitively satisfying picture as applied to adsorption equilibria. Similar compensation effects are frequently found in homologous series of homogeneous and catalytic reactions and there is reason to believe that compensation will also operate in chemisorption. Without stretching credulity, the trend visible in Figure 9.1 allows us to calculate the least squares trend line through the origin as shown.

It is not surprising that there is considerable scatter around the trend line. The compensation effect is expected to operate most precisely in *homologous* series of systems whereas the points shown are a motley collection of data taken from a variety of sources. One can expect that in fact each of the systems (points) plotted will belong to a trend line specific for the family to which that particular point belongs. What we have on Figure 9.1 is a collection of points from such families whose several trend lines obviously lie in a narrow fan in this set of coordinates. There may well be reason for this too. The one commonality one might expect is that all the trend lines would pass near, or even through, the origin. On Figure 9.1 we see that the overall trend line does go through the origin. This agrees with our intuition which says that if the heat of adsorption is zero there is no adsorption, and one can readily expect to find no entropy of adsorption associated with this non-event.

Broadly speaking then, chemisorption equilibrium constants are expected to cluster around the trend line shown in Figure 9.1 and to be limited on the enthalpy side by the strength of the weakest bond in the adsorbing species. Release of more energy than that is likely to lead to bond breakage.

There remains the issue of whether this correlation holds in the region where either or both the enthalpy and the entropy are positive. Although the data is sparse and badly scattered there is no reason to believe that endothermic adsorption is impossible. Endothermic homogeneous reactions are reasonably common and there is no reason to think that such a type of adsorption does not exist. This leads us to suspect that the correlation shown can extend to the right, into the first quadrant. This would imply the possibility of positive entropies of adsorption. Given that the correlation is not perfect, one can also expect that equilibrium constants for low-exothermic equilibria may in some cases show positive entropies, and conversely, that negative entropies could be associated with low values of the entropy of adsorption.

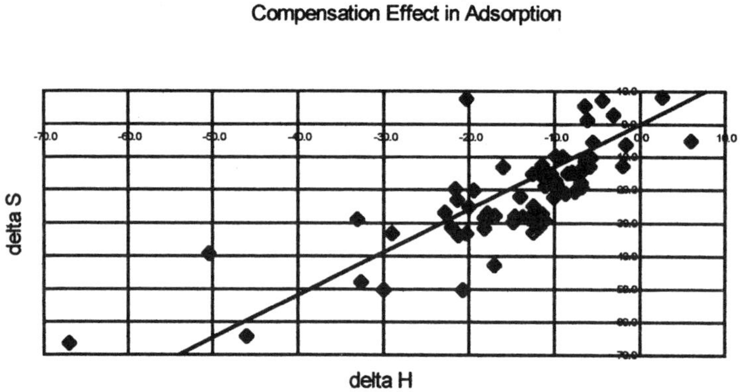

Figure 9.1
Selected entropy-enthalpy values from adsorption equilibrium constants in published rate expressions. There is a surprising lack of well defined values of this type, making it difficult to understand or document the behaviour of adsorption in catalysis.

Conclusions Regarding the Entropy and Energy of Adsorption

Although the subject of limiting parameter values in adsorption equilibria has not been given much (certainly not enough) consideration, it is clear that heats of adsorption are limited on the exothermic side by the strengths of the bonds being formed. On the endothermic side the limit is unknown, although surely it is lower than the strength of the weakest bond in the adsorbate. Otherwise, the energy required for adsorption would be greater than that required for dissociation of the adsorbate and reaction would take place before adsorption occurred.

As for the entropy of adsorption, there appears to exist a compensating effect operating between the enthalpy and entropy of adsorption. The correlation is around a line

$$\Delta S \text{ (entropy units/mol)} = 1.3\Delta H \text{ (kcal/mol)}.$$

Entropy will therefore also be limited in some way, perhaps via the compensation effect by the more easily measured enthalpy considerations. One could look to Figure 9.1 as a guide for identifying whether the entropy and enthalpy of a newly calculated equilibrium constant lie close enough to the visible correlation to be realistic.

Experimental Rate Parameters in Catalytic Reactions

The overall activation energies and frequency factors of overall catalytic rate expressions, and consequently their temperature coefficients, are less easy to constrain. For example, the relatively simple rate expression for the DAM model of the oxidation of carbon monoxide on a platinum on alumina catalyst (see Chapter 11) is thought to be:

$$r_{CO} = \frac{k_r K_{CO} K_{O2}^{1/2} P_{CO} P_{O2}^{1/2}}{(1 + K_{CO} P_{CO} + K_{O2}^{1/2} P_{O2}^{1/2})^2} \tag{9.6}$$

This expression in its full form can reduce to simpler forms and for purposes of illustration we will consider the following:

1. The denominator approaches 1 and the overall rate reduces to:

$$r_{CO} = k_r K_{CO} K_{O2}^{1/2} P_{CO} P_{O2}^{1/2} = k_e P_{CO} P_{O2}^{1/2} \tag{9.7}$$

where k_e is the experimentally observed rate constant. In this example, if the heats of chemisorption for CO and O_2 are exothermic, the overall activation energy will not exceed that possible for the elementary process of conversion of the adsorbed intermediate to products, as governed by the rate constant k_r. In fact, if the heats of adsorption are exothermic and large, the activation energy could become negative, leading to a negative temperature coefficient. On the other hand, if the heats of adsorption are endothermic, all bets are off and the experimental overall activation energy could be much higher than the bond strengths involved.

2. The adsorption term for CO dominates the denominator and the overall rate reduces to:

$$r_{CO} = \frac{k_r K_{O2}^{1/2} P_{O2}^{1/2}}{K_{CO} P_{CO}} = k_e \frac{P_{O2}^{1/2}}{P_{CO}} \tag{9.8}$$

In this example too, similar unexpected sizes and signs of activation energies are possible. The conclusion is that in catalytic reactions there are no readily definable limits to experimental overall activation energies or frequency factors. In order to determine if an experimentally determined overall parameter is acceptable we need to know its structure in terms of adsorption and rate constants from a mechanistic formulation of the full rate expression. Only then can we hope to be able to assign estimated values to the Arrhenius parameters of these more elementary constants. We can then compare the observed values of the composite constant with those that are justified when acceptable parameter values are assigned to the constituent elementary constants.

Anomalies

As a final warning, see Figure 9.2. There the overall rate of reaction according to the above rate expression for CO oxidation is shown. This behaviour occurs only for some select sets of rate parameters. One such set (units are not important in this simulation) is as follows:

$$A_r = 1.41e{+}04; \quad \Delta E_r = -1.94e{+}04;$$
$$A_{CO} = 1.05e{-}11; \quad \Delta E_{CO} = 1.41e{+}05;$$
$$A_{O2} = 1.23e{-}05; \quad \Delta E_{O2} = 6.85e{+}04.$$

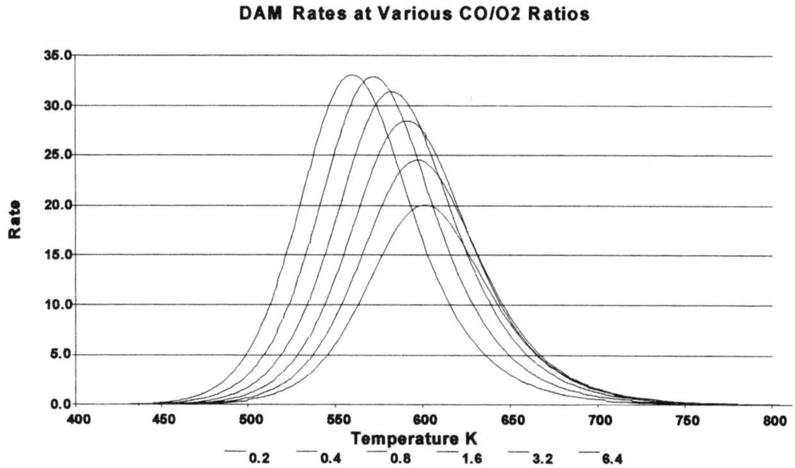

Figure 9.2[3]
Unusual behaviour encountered in the DAM model of CO oxidation at a limited range of kinetic parameters. Over a wide range of CO/O$_2$ feed ratios the rate of reaction increases up to a certain temperature and then proceeds to decrease. This fact, observed over a limited range of conditions, can lead to the misconception that the reaction is exhibiting a negative activation energy when in fact it is simply exhibiting a negative temperature coefficient.

For most sets of parameters the rate expression predicts that the rate continues to increase with temperature. For this set of parameters, however, the overall reaction exhibits a positive temperature coefficient at low temperatures and a negative one at high temperatures. This is so despite the fact that all the individual fundamental parameters are reasonable. Not only does this illustrate the impossibility of proposing limits on overall activation energies, but it clearly shows how confusing overall results can be if one attempts to interpret reaction mechanisms on the basis of overall reaction rates and their parameters.

---

[3] Reprinted from Applied Catalysis, Vol.190A, B.W. Wojciechowski et al., "Kinetic studies using temperature scanning: the oxidation of carbon monoxide," pp.1-24 (2000) with permission from Elsevier Science.

Understanding Rate Parameters

The kinetic parameters obtained by fitting mechanistic rate expressions must be reasonable, in the sense of the above discussion, if the rate expression is to be taken as being representative of the mechanism of the reaction. Since there are numerous cases where identical or similar rate expressions can result from several mechanisms, the appearance of unacceptable parameters can sometimes serve as an indicator that our understanding of the mechanism behind the rate expression is in error. At that point the search for the correct mechanism must be redirected to finding a mechanism that leads to a similar form of the overall rate expression but involves a different interpretation of the rate parameters in the overall rate expression.

> *The process of iterating from raw data through a series of data handling procedures and fitting routines ends on a final question: is the interpretation in agreement with possible physical realities? If not, other interpretations have to be sought, perhaps after expanding the experimental work or by reconsidering the available data.*

It may even be necessary to select an interpretation that is less than the best fit in terms of statistical measures of fit, a point already discussed at the end of Chapter 8. If, on the other hand, the interpretation is statistically optimal and agrees with physical realities, one must expect this conclusion to be subject to independent tests, tests not connected with kinetic studies that led to the original conclusion. In all these cases the judgment of the interpreter has its place as a proper influence in the interpretation and is indispensable to the construction and verification of alternative postulates.

10. Statistical Evaluation of Multiparameter Fits

Statistics is a tool for deciphering information contained in experimental data. Like any tool, it can be used to produce things of lasting value, or merely to reshape the raw material. It is the skill and insight of the user that makes the difference.

Introduction

Statistical evaluation methods are well defined in the case of linear dependencies:

$$y = ax+b.$$

The widely known and widely used methods of linear regression make available the two parameters, a and b, as well as their confidence limits, based on some criterion of probability that the correct parameters will lie within the stated +/– limits of the best-fit value found. In general the regression is based on the method of least squares of residuals (SSR) where the SSR for a straight line fit is minimized.

Confidence limits are derived using the t-distribution setting a desired confidence level, usually 95% or 99%, with (n–2) degrees of freedom, where n is the number of experimental readings. The limits calculated in this way are fairly unambiguous, or at least broadly accepted, and can be used to compare the quality of the parameters obtained in various studies. This kind of treatment appears in sources ranging from elementary statistics textbooks to built-in routines in commonly available spreadsheets. The procedure is so commonly applied that most users of the built-in linear regression routines are probably unfamiliar with the nuts and bolts of the methodology.

This familiar procedure, and in particular the availability of confidence limits from its application, has created the expectation that comparable values can be reported in all fittings, regardless of the complexity of the governing dependencies or the extent of parameter correlation in a given case. The search for a means to make such quantities universally available has led to a number of attempts (see Bates and Watts (1988)) to calculate error limits in non-linear, multi-parameter cases, ranging from the simplistic to the highly idiosyncratic.

The fact is that this is a difficult problem, one which has no commonly accepted solution that applies in all of the many and varied cases where non-linear equations in multiple dimensions need to be fitted. Parameter estimates in such cases are invariably best done in precisely the most non-linear regions of the pertinent variable space (see Chapter 11), where linear approximations do not hold. Several methods have been put forward for both the selection of fitting criteria and the calculation of confidence limits in various instances. One sophisticated method involves the calculation of a matrix of as many rows as there are data points, whose columns are partial derivatives of the fitted function, parameter by parameter. The simplest assumes linear behaviour near the optimum parameters. This variety of choices means that, if confidence limits are to be reported, one must chose one of several available methods of calculating these values.

The very fact that several methods of calculating such limits exist diminishes their utility for comparison purposes and casts doubt on the meaning of the absolute values of the limits.

The lack of a commonly accepted and thoroughly objective method of evaluating multiparameter fits leads to the proposal that *parity plots* be used as the simplest, best, most objective, and undeniably unifying method of determining the "goodness" of fit in such instances. These plots are closely connected with *deviation plots,* and together the two presentations are a valuable guide to the determination of best fit in all multiparameter systems. The procedure described below affords both a visual and a numerical criterion of fit which is independent of parameter correlation, variable space curvature and operator judgment. It also allows ready comparisons of data-sets from different laboratories and various regions of variable space.

The Parity Plot

A parity plot can be constructed for the fittings of dependencies involving fits using any number of variables. The resultant plot can be compared to plots of corresponding data obtained in other laboratories or at other times. Since all parity plots should in principle look similar, it is even possible to compare the quality of fits between systems that have nothing in common. This property allows us to judge the quality of fits in various systems, the scatter in data obtained using different experimental setups, or the quality of work issuing from several sources. As a quality control tool it is incomparably better than most other measures which are constrained by the complexity of the systems under consideration and hence tend to lack universality. Since the parity plot is inherently linear, the SSR and the accompanying confidence limits of a linear regression of this plot can be calculated to quantify scatter about the line of parity.

Constructing a Parity Plot

A parity plot can be constructed for any data set, at any stage of its processing. For example, one can draw a parity plot for the raw data against a cleaned-up (i.e. filtered or mass balanced) version of the same data (see Chapter 7). Most commonly it will be used to evaluate the final fit of calculated values against the raw data, or data that has been cleaned up in some way. Any two compatible sets of data can be examined to see the differences between them.

The parity plots we will concentrate on are a graphic representation of experimental data plotted against its calculated value from the fitted equation designed to represent the system. Either quantity can be plotted on either axis but, in the interest of consistency, we will adopt the convention that the experimental value is plotted on the x-axis while the corresponding calculated value is plotted on the y-axis. Each point therefore has the co-ordinates (Experimental Value, Calculated Value).

A perfect fit between experimental values and values calculated using a fitted expression will result in a 45° straight line in the positive-positive quadrant of the plot. This is the *line of parity*. If there is noise in the data but the fit is satisfactory overall, there will be a scattering of points about this line. The SSR is a valid and useful measure of the scatter about the parity plot.

Evaluating the Goodness of Fit Using a Parity Plot

A quantitative measure of goodness of fit can be obtained by performing a linear regression on the data assembled in the parity plot and observing the following:

- the confidence limits on the two parameters, slope and intercept, reported from the linear fitting;
- whether the intercept is zero, as it should be in principle.

What constitutes acceptable numbers in each case is up to the user to judge. However, these numbers are strictly comparable with corresponding numbers from other data-sets or from parts of the same data-set that might have been generated under several different conditions or that might be otherwise distinguishable. Regardless of the complexity of the relationships underlying the data, such a plot will result when data is plotted against fitted values.

A visual inspection of the parity plot will usually reveal any significant deviations from the line of parity. These must be examined in more detail. There are a number of types of deviation that can serve to direct attention to specific problems with the fit. For example: there may be missing terms in the correlating model equation; the analytical form of the proposed model may be completely unsatisfactory; or the fitting routine may need to be restarted with different parameter values to take account of a significant misfit that is not being properly addressed by the fitting subroutine. We will examine several of these in turn.

Deviations from the Line of Parity

Misfits can be made more apparent by plotting the deviation of the experimental value from the parity line on the y-axis against the experimental value itself on the x-axis. Such plots should be generated routinely for the overall experimental data-set or for appropriate subsets of data. In general, two types of phenomena are observed in such deviation plots.

The scatter is uniform about the x-axis. In this case there is no trend in the scatter and the sum of the deviations tends to zero. A linear regression of this data gives zero intercept and zero slope for the deviation plot. This behaviour identifies random *noise* as the cause of the scatter. Trends in the scatter become apparent and may well differ between distinguishable sets of data from the overall parity plot. The trends seen can be monotonic or linear or even weave around an overall trend. This behaviour identifies *distortion* as the cause of the misfit.

These two behaviours distinguish two very different causes of misfit between the data and its fitted equation. Noise is an unavoidable consequence of dealing with physical realities, while distortion is an inadequacy of the fit.

The meaning of noise

Noise is the result of random error due to control input/output functions, errors in analysis, digital dither in the electronics, and a potential host of presumably random causes. The noise level may be constant, or may vary over the range of data gathered. In either case, the sum of deviations cancels out in each increment of the range. Although these subtleties are important, they are often left unexamined in fitting analytical expressions to experimental data.

An illustration of the consequences of this neglect can be seen in the commonly used approach to data fitting, when commercial routines that adjust the parameters of the fitted equation to achieve a satisfactory fit are used. For these purposes deviations are commonly quantified in the form of the sum of squares of residuals (deviations), the SSR, between the measured and the calculated values by taking:

$$SSR = \Sigma_{\text{all measurements}}(\text{fitted value} - \text{measured value})^2$$

The relationship between physical realities and fitted parameters of mechanistic rate expressions is very important. Realistic and hopefully correct parameters tell the investigator something about the thermodynamics and binding properties of the species involved. Unrealistic parameters are of no use in such considerations, although they may well provide values that can be used in modeling the system for purposes of reactor design or process control. These two uses of fitted rate equations should be clearly distinguished and even reported as such when the results are presented.

The single-criterion procedure, using the SSR, obviously neglects all meaning that might be found in changes in patterns of scatter over the range of the experimental measurements. If the scatter is more pronounced in one region of measurements its influence will, rightly or wrongly, be greater on the fitting than that from less noisy regions. Moreover, the SSR of the final fit says nothing about the adequacy of fitting over the range of measurements.

The total deviation can also be quantified as the sum of deviations (SUD). For a perfect fit this number should be zero. Other measures of deviation may also be appropriate, as we will see. However, minimization of the SSR is normally used as a criterion in the curve fitting routines available for this purpose. This may be a good way to find the region where the optimum solution will lie but not necessarily the final criterion for this purpose.

Figure 10.1 shows the fitting of a rate data set for carbon monoxide oxidation to the expression:

$$r_{CO} = \frac{k_r K_{CO} K_{O_2}^{1/2} P_{CO} P_{O_2}^{1/2}}{(1 + K_{CO} P_{CO} + K_{O_2}^{1/2} P_{O_2}^{1/2})^2}$$

where the constants have the usual Arrhenius dependencies. The same data is discussed in Chapter 11. The best fit is characterized by:

$$SSR = 9.81E{-}02; \quad SUD = -3.29E{+}00$$

The parity plot in Figure 10.1 shows a very good fit that on closer examination of a deviation plot is seen to be less satisfactory. A second pass at fitting the same data was made using the parameters from the SSR search as starting parameters for minimizing the SUD. The results are shown in Figure 10.2. The corresponding statistical values are:

$$SSR = 1.25E{-}01; \quad SUD = 1.01E{-}07$$

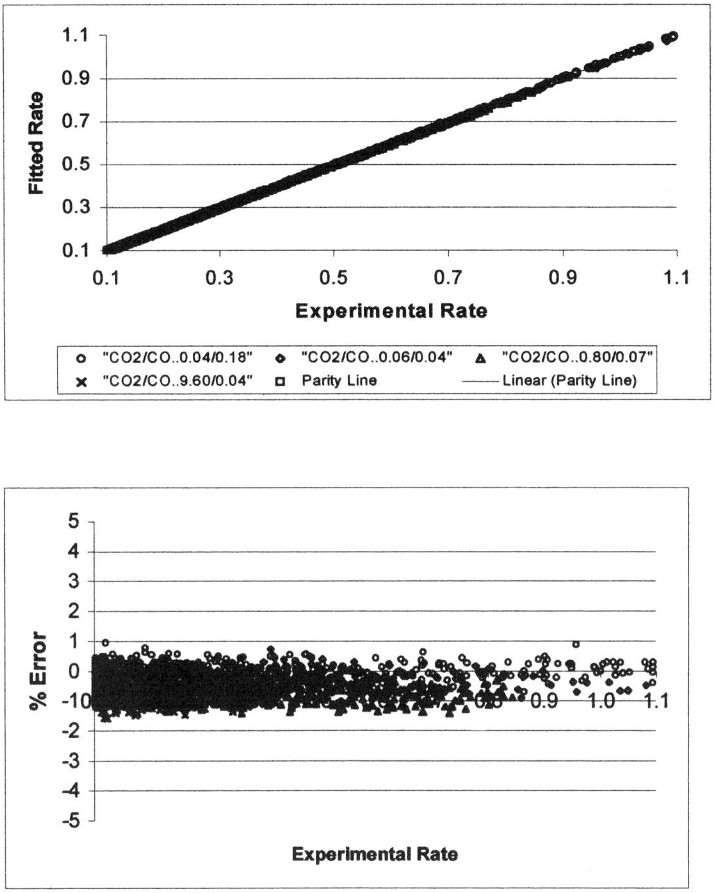

Figure 10.1

The parity plot of a fit obtained using the SSR as a criterion of fit is shown in the first part of this figure. The corresponding deviation plot shows that errors are not evenly distributed about zero and distortion seems to be present in the several of the reactant compositions tested.

Which fit is better? Figure 10.2 clearly shows less distortion on the deviation plot even though the SSR for this fit is higher. Often improvements in fit can be obtained by fitting with restricted sets of the variables, after the global fit using all the variables has settled into a minimum. This and other procedures are given at the end of the chapter.

It can be argued that both solutions are acceptable if the SSR lies within some range deemed to be adequately small. In this example the parameters do indeed differ insignificantly between the two fits. This need not be the case in all instances and the fitter is then free to choose whichever adequate set of parameters suits some other, external, criteria. To some this may constitute a dilemma, since the fitter is picking a preferred set of parameters over other equally good, or even better, sets. It should not be.

Complete objectivity in parameter determination cannot be obtained solely by using statistical measures of fit of complex models.

> *In chemical kinetics in particular, guidance in selecting an acceptable set of rate parameters* ***must*** *be sought in the physical meaning and constraints on the parameters found and in information obtained from auxiliary considerations such as thermodynamics, adsorption studies, etc.*

Sight of this compelling fact has been lost in the fog of statistical numbers that are routinely reported to justify the validity of the parameters obtained.

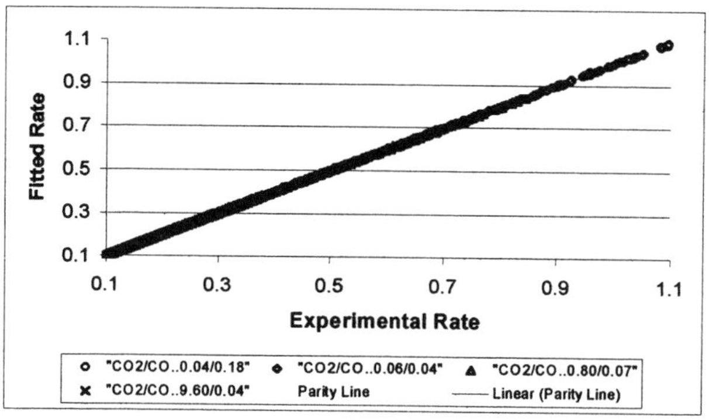

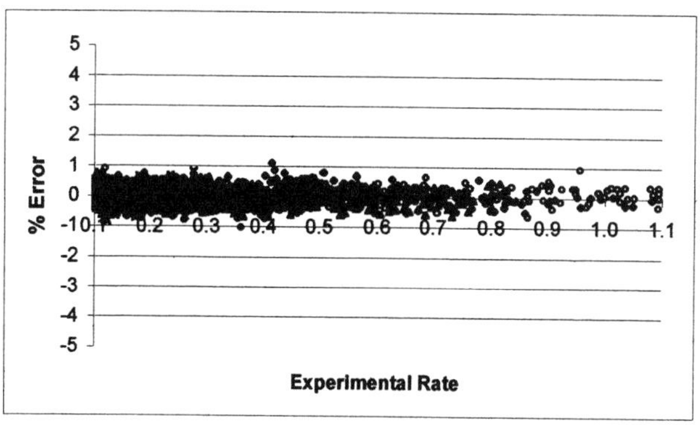

Figure 10.2
Using an alternative fitting criterion based on the sum of deviations, SUD, but starting with the fit in Figure 10.1, results in a better fit as evidenced by the above deviation plot. In this particular case the two fits are so good that the difference between the resultant parameter sets is in the third place of decimals. However, there is no reason to believe that such differences are always negligible and that meaningful differences in alternative fittings cannot arise.

Order of fit

There is reason to believe that other very simple criteria of fit are better in some cases. Consider the following *criterion of fit* (COF):

$$\mathrm{COF} = \sum_{\mathrm{all}} (\mathrm{SR})^{n}$$

where SR is a square of the residual between an experimental and a calculated value.

The n indicates the exponent applied to each squared residual (SR) to be included in the sum of the COF. It is not the *sum* of squared residuals that is raised to the power n. This criterion is much more flexible than the simple SSR and allows the fitter to vary the criterion of fit by varying n, the order of fit, to suit the circumstances. Figure 10.3 shows some consequences of adjusting n.

Assume over the range of measured values from 1 to 40 there is a distribution of deviations between the fitted value and the measured value, lying between 1 and –1. For purposes of illustration we take it that the errors lie along a line as shown by the triangles in Figure 10.3. The real distribution need not lie along this line but can take any form, or be randomly scattered, without negating the arguments below.

In order to get a global fit one must select some measure of misfit pertinent to the whole range of conditions studied. For this purpose we require positive numbers only to describe the errors so that we do not arrive at a bad fit whose distribution of (possibly large) errors sums up to zero. For example, the sum of deviations expressed by the triangles is zero and using SUD would tell us that the fit with this set of errors is perfect. To avoid this misconception one normally uses a sum of the squares of deviations (residuals), the SSR. The fitting is done by reducing this sum. This is the procedure for obtaining a tight fit about the line of parity using the common linear regression methods. However, the squares of error, shown as circles in Figure 10.3, give a different emphasis to error in different regions than that which would be most even-handed in view of the actual errors displayed as the line of triangles in Figure 10.3.

This raises an important side issue involving the absolute size of the errors. In order to bring these values to a common basis it is often best to use fractional deviation between the fitted and the measured value.

$$\text{Error} = (\text{Fitted value} - \text{Measured value}) / (\text{Measured value})$$

A fractional deviation defined on the measured value will generally result in errors lying between 1 and zero. Errors larger than 1 will still be properly processed, but might well give pause to the experimentalist, and the salient point remains: squares of deviations distort the weights of deviations according to size and have a consequent effect on the fit using the SSR fitting criterion. In essence they weight the deviations, biasing their influence on the fit.

The simplest remedy is to use $n = 0.5$ as the order of fit. This results in the errors being represented by the diamonds in Figure 10.3. Clearly, errors are now fairly represented vis-á-vis their real weight, with the added benefit that all errors are represented by positive numbers. This allows us to minimize an undistorted COF.

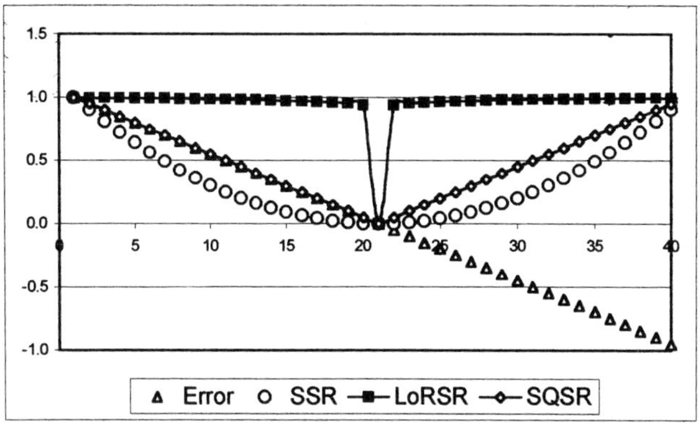

Figure 10.3
Variations in the value of individual point misfit with the order of fit. Triangles represent original error; circles represent the square of error (n=1); squares, a very low (n=0.01) order of fit; and diamonds represent the square roots of the squares of error (n=0.5)

There are benefits, however, to the use of a distorted COF. The squares on Figure 10.3 show one form of useful distortion. There, all errors are given almost the same weight by using a very low value of the order of fit: n = 0.01 in this case. The result is that all errors are given equal attention by the fitting algorithm, but, more importantly, the *surface of errors* that is searched for minima by the fitting algorithm is different from that for the simple SSR or, for that matter, for any other value of the order of fit.

Changing the order of fit after one sinks into a "best solution" on one fitting surface can offer a chance to improve the fit, as judged by the fitter, by joggling the parameters to new values. In this way one might escape a local minimum and proceed to a better solution, whatever that may be, using the alternative COF. This procedure will only work for imperfect data; otherwise, all COFs would result in the same parameters and zero for the COF. Fortunately imperfect data is easy to get and real data has to be massaged to achieve a final result (see Chapter 7) that is both acceptable statistically and likely physically.

Here we come to the heart of the matter: is there an obviously "best" COF? Opinions may vary but it seems that, for pure noise, the COF using n = 0.5 might be preferred. The final decision whether the parameters obtained using any COF are acceptable for theoretical work is made on the basis of auxiliary considerations, such as those mentioned above and in Chapter 9. On the other hand, the best fit using unconstrained parameters is normally the solution sought when empirical fitting is designed to reproduce the behaviour of a system for purely practical purposes. Such parameters require a minimization of the COF using all the methods available until a truly best fit is obtained.

The meaning of distortion

Distortion is very different from noise and can arise from one or several of the following:

- the use of an improper modeling function(s) to represent the data;
- failure of the fitting routine to find the global minimum in fitting the data;
- the existence of a systematic error due to calibration or some other unrecognized bias in the data.

Noise is normally superimposed on distortion, making the deviations from the line of parity fuzzy. Because by definition the form and/or extent of the distortion differ along the range of measurements, distortion can sometimes be used as a qualitative guide to improve our understanding of the reasons behind the misfit, thereby helping to direct the search for the best fit equation for the data. How to interpret the causes of the distortion and use this as a guide to further data fitting depends very much on the functional dependencies in the equation(s) used to describe the data.

The clearest sign that distortion is present is seen when the noise about the distorted trend is small relative to the distortion. Distortion then clearly indicates that a correlation exists in the data *but it is not the correlation being attempted.* This reading of the deviation plot offers an opportunity to use the information thus supplied to improve the fitting criterion in searching for the best parameters. The conventional numerical SSR criterion ignores the form of the deviation revealed on the deviation plot and cannot satisfactorily deal with distortion.

Figure 10.4 shows an example of distortion in fitting the same data as that in Figures 10.1 and 10.2. Convergence on the line of parity is excellent but the deviation plot clearly shows that deviations exist and differ for the several reaction condition sets used. The statistical fit as measured by the SSR and SUD is not bad when compared with the results reported above.

$$\text{SSR} = 2.44\text{E}{-}01; \quad \text{SUD} = 4.17\text{E}{-}00$$

One set of results, that represented by circles, is actually fitted very well, except for an easily removable offset, and only the results of runs under other experimental conditions show that the model or the parameters are inadequate. The model used is the same for Figures 10.1, 10.2 and 10.4 and in Chapter 11. However, the parameters for Figure 10.4 are vastly different than those for Figures 10.1 and 10.2. In fact, whereas the exponents of the adsorption equilibrium constants are positive in 10.1 and 10.2, they are negative in 10.4. The original authors rejected the inferred enthalpies and gave no further consideration to this solution, on the assumption that the pertinent adsorptions of oxygen atoms and carbon monoxide molecules are exothermic.

The above methods are known, but under-utilized in practice. More on this subject, from a statisticians point of view, can be found in a book by Draper and Smith (1981).

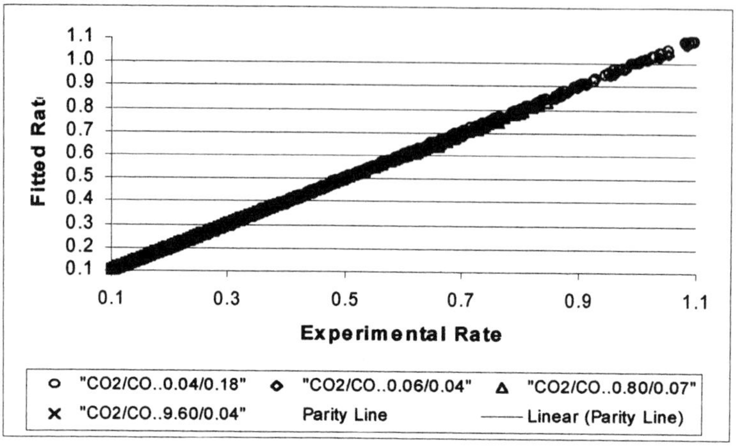

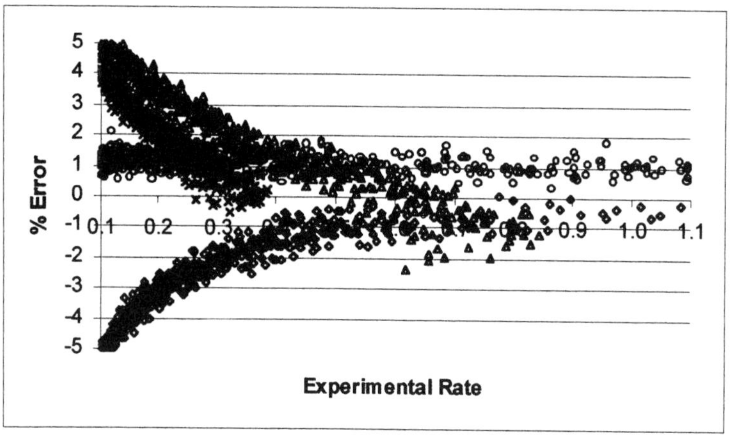

Figure 10.4
A fit showing good convergence on the line of parity but significant distortion when viewed on the deviation plot.

A List of Suggested Procedures for Data Fitting

In fitting data with analytical algebraic expressions, one normally finds that the available fitting algorithms do not zero in on the desired fit without substantial inputs from the fitter. Perhaps one of the most important choices one must make to mitigate this problem is the selection of a fast and appropriate fitting algorithm (see Chapter 8). That done, one proceeds to refine the fitting of a candidate rate expression. Below is a list of procedures that can be useful while searching for the desired fit and its parameters.

1. Establish a good set of initial parameters. In fitting rate expressions using TSR data, this can be done by fitting several (as many as is convenient) isothermal data sets with the proposed rate expression and plotting the resultant constant temperature parameters on Arrhenius plots (see Chapter 8). The best estimates of the Arrhenius parameters of each rate parameter obtained in this way are then used to start the iterations for an all-up fit. Notice that if the Arrhenius plots for each of the temperature dependent rate parameters are not linear, the model being used in the isothermal fitting is inadequate (see Chapter 9).
2. Iterate to a minimum using the familiar SSR criterion. Make sure that iterations do not terminate due to a limit on the number of iterations rather than a limit set on the desired quality of fit.
3. Change the COF and re-iterate. If the solution is better, keep it. If not, change the COF again and repeat.
4. Go back to any previously used COF to see if the new starting parameters can lead to an improved fit.
5. Change the COF, alternating between the SSR of the absolute values of deviation and that of their percentages. Start with the SSR then go to 3.
6. Change fitting algorithm procedures, such as an alternative use of the derivatives, an alternative search procedure, etc. Availability of these features will vary with the algorithm used.
7. Tighten the criterion of fit that terminates iterations of the fitting algorithm.
8. Try fitting subsets of the best parameters determined. For example, fit only the pre-exponentials, or only the energy terms, or only the terms in the denominator.
9. Try other algorithms.
10. If no satisfactory fit is obtained, re-examine the Arrhenius plots of the data using the best parameters established to date as the initial parameter values for the fitting. If any of the plots fail to show good linearity and show distortion rather than pure noise, either the data or the model is at fault.
11. Check the data in its processed form against the raw data to see if distortion has been introduced by any preceding manipulation of the data.

The list is not exhaustive, and practitioners will find other procedures that can be helpful. Time can be saved by automating several of the above steps to execute in a predetermined order, using a macro or a subroutine. Eventually an acceptable model and fit will be obtained. This raises the final question. What makes this fit acceptable?

11. Experimental Studies Using TSR Methods

The development of a new experimental technique has often resulted in the rejuvenation of an old field of research, or the opening of a new field of enquiry. The method of temperature scanning for collecting reaction rate data holds out a promise of restoring the utility of reaction rate studies in understanding chemical processes.

Applications of Temperature Scanning Reactors

Temperature scanning methods have been successfully applied in university and industrial research to a number of studies of the kinetics of chemical reactions. The simple, homogeneous, liquid phase hydrolysis of acetic anhydride has been studied in a TS-BR, a TS-PFR and a TS-CSTR. The results are available in the literature (Asprey, S.P. et al., (1996)). A study of the catalytic oxidation of carbon monoxide using platinum on alumina (Wojciechowski, B.W. and Asprey, S.P., (2000)) was done using a catalytic TS-PFR. A study of the steam reforming of methanol (Asprey, S.P. et al., (1999)) broadened the range of complexities tackled using this method. Figure 11.1 shows the configuration of the TS-PFR used in these catalytic studies.

Figure 11.1
The first TS-PFR. The scale of the setup is given by the two 14" monitors on the tables beside the TS-PFR while the level of automation is indicated by the fact that only four power switches on the front panel require manual operation.

A comparison between a conventional and a TS-PFR study of methanol reforming is contained in the paper by Asprey et al. (1999) and the associated paper by Peppley (1999). Other workers have used gas phase TS-PFRs in a number of studies carried out in industry. An example of industrial work is given in "Investigation of the Kinetics of Ethylbenzene Pyrolysis Using a Temperature Scanning Reactor", Domke et al. (2001). Below we present some of the results and observations from selected studies and relate them to the issues raised above.

The Oxidation of Carbon Monoxide

Temperature scanning (TS) made it possible to complete an experimental study of the kinetics of this reaction on one catalyst, at one pressure and feed composition, in less than one working day of reactor operation. Real-time measurements of CO conversion were done using a quadrupole mass spectrometer.

The Experimental Data

The kinetics were quantified using >12,000 rate - conversion - temperature (r, X, T) triplets calculated from data obtained using a TS-PFR. Four experiments at various feed compositions were used to improve parameter estimation. After rejecting a number of possible rate expressions and their mechanisms, the results from the TS-PFR were fitted with the remaining two candidate rate equations based on mechanistic considerations:

- the Langmuir-Hinshelwood dual site Molecular Adsorption Model (MAM);
- the Langmuir-Hinshelwood dual site Dissociative Adsorption Model (DAM).

The two models differ in their view of the state of the adsorbed oxygen:

- the MAM model presumes a reaction of adsorbed carbon monoxide *molecules* with adsorbed *molecular* oxygen, adsorbed on the same type of site;
- DAM model involves the reaction of adsorbed oxygen *atoms* with adsorbed carbon monoxide *molecules*, both adsorbed on the same type of site.

Other models, considered at an earlier stage of investigation, were easily rejected. It also became apparent soon after the four full data sets were examined that the fit to the MAM model was consistently less satisfactory than that to the DAM model. This conclusion could be drawn well before a satisfactory final parameter fit of the DAM model was obtained. As a result the MAM model was also abandoned, allowing for a significant reduction in data fitting effort in the final stages of interpretation.

In general during the course of a temperature-scanning experiment, four measured quantities are of primary interest:

- space time, τ, (measured using mass flow meters);
- reactor inlet temperature T_i, (measured using a thermocouple);
- reactor exit temperature, T_e, (measured using a thermocouple);
- outlet composition, $X_{i,out}$, (measured using a mass spectrometer)

All variables except the space time are functions of clock time t, the time since the start of a given ramp.

Ten or so runs (rampings) were carried out in each experiment. Ramping rates in each run were identical, as required by the procedures outlined in Chapter 4. The collected data was collated in files in preparation for the series of manipulations required on the way to rate extraction (see Chapter 7). First the data was corrected for atomic balances as described in Chapter 7. The treated data was then filtered to smooth the kinetic surface.

Next, the mass of treated data was sieved to form the desired number of tau-conversion-temperature (τ, X, T_e) triplets of data on preselected operating lines. Each triplet on an operating line is for the same clock time, t, (and therefore the same inlet temperature, T_i). These data triplets are then used to extract rates and examine the quality of the data collected (see Chapters 7 and 8).

Figures 11.2 and 11.3 show the raw conversion and reactor exit temperatures as functions of clock time. This is the raw data that might be used for the triplet formation. The experimental conditions used for this data-set are:

- 6 vol. % CO; 4 vol. % O_2; balance Ar,
- temperature ramping from 350K to 550 K,
- temperature ramping rate of 5 K/min, (feed and reactor surroundings),
- conversions from 0 to 100%.

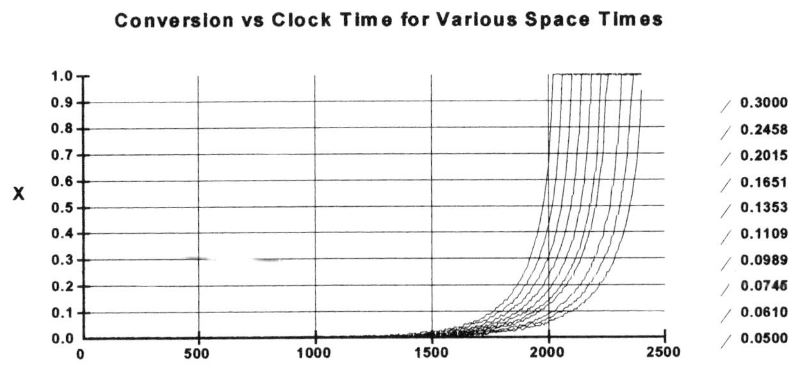

Figure 11.2[4]

Conversion versus clock time at various space times (taus) in a TS-PFR experiment on the oxidation of carbon monoxide. The steepness of the curves at high conversion attests to the high rates of reaction compared to those at short clock times, when the feed temperatures are low. The reaction is essentially irreversible and proceeds to 100% conversion for the feed composition used. Taus are shown on the right. The shortest tau refers to the rightmost curve.

[4] Reprinted from Applied Catalysis, Vol.190A, B.W. Wojciechowski et al., "Kinetic studies using temperature scanning: the oxidation of carbon monoxide," pp.1-24 (2000) with permission from Elsevier Science.

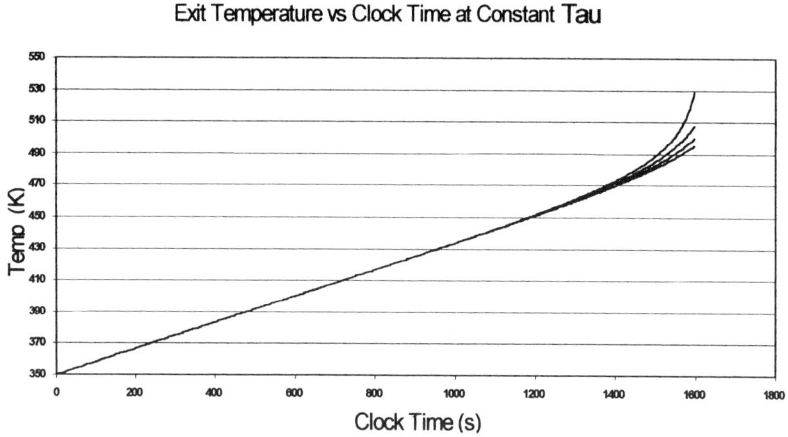

Figure 11.3[5]
Output temperatures during the TS-PFR experiment on the oxidation of carbon monoxide. Due to the configuration of the reactor it was possible to maintain almost isothermal conditions during each ramping. Output temperatures deviated from the ramp temperature only at high conversions, where the heat of reaction could not be fully removed from the reactor.

Three other data-sets, at different feed composition ratios, were collected and treated in the same manner. The same conversion and reactor exit temperature data can be re-mapped onto the X-T_e plane, as shown in Figure 11.4. Figures 11.2, 11.3 and 11.4 represent a large volume of kinetic data if we allow for interpolation. Such interpolation is an integral part of TS-PFR data interpretation and is made more reliable by the frequent sampling for analysis and by the filtering of this raw data, as we will soon see.

It is not hard to see that the curves on Figure 11.4 are actually the two dimensional projections of specific traverses on a more general three-dimensional surface in the coordinates (τ, X, T_e(or t)). In fact, it is easier to grasp the vast amount of data available from a TS-PFR experiment by constructing a 3D surface from the same data as that in Figure 11.3, as shown in Figure 11.5.

In Figure 11.5 we see that the raw data, corrected or not, can create a wrinkled surface. This is due to experimental error, or noise, present in the raw data. Filtering this data using procedures described in Chapter 7 makes it possible to obtain a smooth (τ, X, t) surface and thence a smooth (τ, X, T_e) surface. By taking appropriate slopes from this latter surface, as described in Chapter 5, we obtain a smooth (r, X, T_e) surface, From this surface we can read off any point within the area covered by the experimental data. We read off as many points as we wish and "sieve out" from this collection numerous sets of isothermal (r, X, $T_{e(const)}$) triplets. As many such sets as we feel are needed are then used for conventional fitting to isothermal versions of the candidate rate expressions.

---

[5] Reprinted from Applied Catalysis, Vol.190A, B.W. Wojciechowski et al., "Kinetic studies using temperature scanning: the oxidation of carbon monoxide," pp.1-24 (2000) with permission from Elsevier Science.

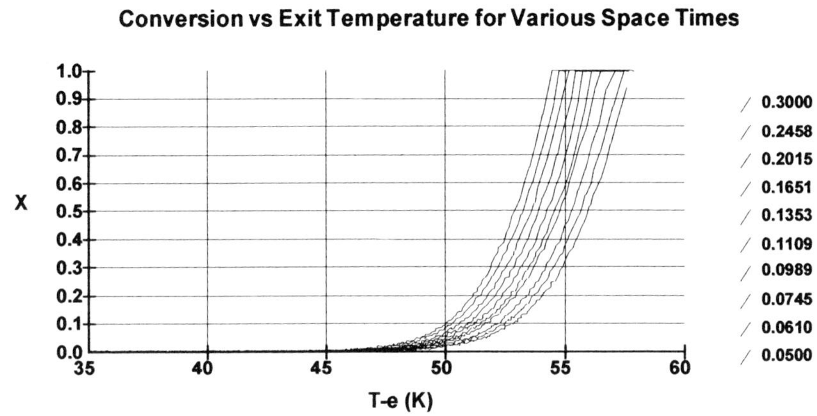

Figure 11.4[6]
The same results as those shown in Figure 11.2 are shown on the reaction phase plane after remapping using the temperature trajectories shown in Figure 11.3.

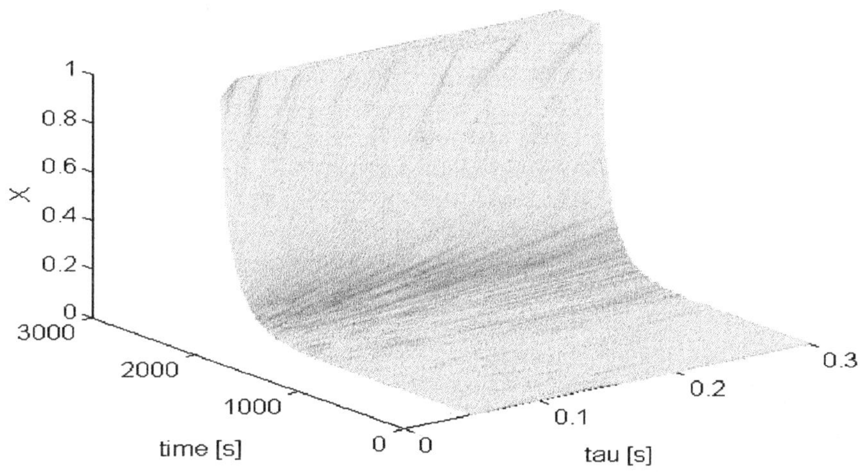

Figure 11.5[6]
Surface of unfiltered results such as those shown in Figure 11.2. This conversion surface has to be remapped as an X-T_e-τ surface for determining reaction rates. The wrinkles visible induce unacceptable scatter into the slopes and must therefore be removed using mathematical filters.

---

[6] Reprinted from Applied Catalysis, Vol.190A, B.W. Wojciechowski et al., "Kinetic studies using temperature scanning: the oxidation of carbon monoxide," pp.1-24 (2000) with permission from Elsevier Science.

Three sets of data, sieved to yield constant-temperature triplets, were used to obtain the initial estimates of the parameters (kinetic constants) in these rate expressions. Arrhenius plots of these constants yielded the initial estimates of the activation energy and frequency factor of each of the temperature-dependent parameters. The initial values of the Arrhenius parameters were then inserted into the "all-up" rate expressions, including all the Arrhenius parameters.

For this second stage of fitting, a large number of (X, r, $T_{e(const)}$) points are chosen, at many temperatures, and broadly distributed over the available rate surface. The complete rate expression, with the Arrhenius parameters of each constant included, is then fitted to all these points simultaneously. Optimization, as described in Chapter 7, yields the best-fit values of the parameters of this expression. Several attempts to reach the same optimum set of parameters are made, with somewhat different starting values, to make sure that a stable and optimal solution has been found.

Fitting the Data

In the case of the oxidation of carbon monoxide, the mechanistic rate expressions investigated were of the general form:

$$-r_{CO} = -\frac{dP_{CO}}{d\tau} = \frac{k_r K_{CO} K_{O2}^m P_{CO} P_{O2}^m}{(1 + K_{CO} P_{CO} + K_{O2}^m P_{O2}^m)^n} \tag{11.1}$$

where the P_i is the partial pressure of component i in the output of the reactor.

Fits of two principal reaction mechanisms, both of which have the above general form, were made, after initial trials of rate expressions corresponding to mechanisms with other forms of rate expression had resulted in the rejection of these forms. In the above equation the Molecular Adsorption Model (MAM) predicts n=2, m=1 while the Dissociative Adsorption Model (DAM) leads to n=2, m=1/2. The two mechanisms differ in that MAM assumes that adsorbed molecular oxygen reacts with adsorbed carbon monoxide molecules, both of which reside on identical sites. Alternatively, the DAM assumes that the adsorbed oxygen molecules dissociate into atoms before reaction with the adsorbed carbon monoxide molecules, once more both residing on identical sites. The two concentration exponents, referred to as orders of reaction, are temperature independent and integral. All the other constants are temperature dependent and follow the Arrhenius relationship. These comprise k_r, a catalytic rate constant, and two adsorption equilibrium constants K_i, all subject to the constraints described in Chapter 9. Notice that a mechanistic rate expression always presumes that the rate is measured at constant volume.

Since the experimental data are reported in mole fractions, this mechanistically derived equation, which has been formulated in terms of partial pressures (or concentrations), has to be transformed into a form suitable for fitting to the available TS-PFR data. We do this in several steps following the procedures laid out in Chapter 7.

The first change in variables involves replacing the partial pressures in the mechanistic formulation with the output mol fractions, corrected and/or filtered or neither, depending on the procedures used. We will write them in terms of the corrected mol fractions available in the y vector which was calculated in Chapter 7. This replaces each partial pressure P_i with the corresponding expression in terms of total pressure and mol fraction, Py_i. In the case of the DAM model this gives:

$$-r_{CO} = -\frac{dy_{CO,out}}{d\tau} = \frac{k_r P^{-1} K_{CO} K_{O2}^{1/2}[Py_{CO,out}][Py_{O2,out}]^{1/2}}{(1 + K_{CO}[Py_{CO,out}] + K_{O2}^{1/2}[Py_{O2,out}]^{1/2})^2} \qquad (11.2)$$

where P is the total pressure at which the experiment was carried out.

This creates a generalized expression for the rate of reaction carried out at any pressure P but still at a constant volume. The expression can be used as it stands for constant volume reactions. Since volume contraction occurs in this reaction, we next have to account for it. In this connection we are not so much interested in the change in the output mol fraction of CO (y_{CO}) as we are in the fraction of moles at the input that has been converted. We introduce this by rewriting the rate in terms of x_{CO}, the fraction of the original CO converted (see equation 7.60). This, after some rearrangement, leads to the equation:

$$-r_{CO} = \frac{dx_{CO,out}}{d\tau} = \frac{\dfrac{\delta}{Py_{CO,in}} k_r [PK_{CO}][PK_{O2}]^{1/2} y_{CO,out} y_{O2,out}^{1/2}}{(1 + [PK_{CO}]y_{CO,out} + [PK_{O2}]^{1/2} y_{O2}^{1/2})^2} \qquad (11.3)$$

In treating TSR data, only the left side of the equation needs to be put in terms of fractional yields or conversion $x_{i,out}$. The rate of conversion of CO given by the right side can remain in terms of output mol fractions modified by the transposed volume expansion factor δ from the left and used for fitting the parameters. Note that if there is no volume expansion and δ = 1 we can move the term $Py_{CO,in}$ to the left side, where the rate becomes:

$$r_{CO} = \frac{d(P[y_{CO,in} - y_{CO,out}])}{d\tau} = -\frac{dPy_{CO,out}}{d\tau} \qquad (11.4)$$

In all cases the units and value of the term $\delta/Py_{CO,in}$ are included in A_r. There are therefore several ways that a rate can be expressed. These lead to different values and units of the frequency factor A_r along the lines discussed in Chapter 9 and illustrated in a following section of this chapter. After due account is taken of the units the caveats regarding admissible magnitudes discussed in Chapter 9 should be observed. The space velocity τ is defined here as bulk volume of catalyst V divided by the volumetric flow rate of total feed measured at STP conditions.

Including the Arrhenius parameters in all the pertinent constants results in:

$$\frac{dx_{CO}}{d\tau} = \frac{\left[\dfrac{\delta A_r}{P y_{CO,in}}\right](\Delta E_r/RT)(PA_{CO})(\Delta E_{CO}/RT)(PA_{O2})^{1/2}(\Delta E_{O2}/2RT) y_{CO,out} y_{O2,out}^{1/2}}{(1+(PA_{CO})(\Delta E_{CO}/RT) y_{CO,out} + (PA_{O2})^{1/2}(\Delta E_{O2}/2RT) y_{O2,out}^{1/2})^2} \qquad (11.5)$$

This is the form that is fitted to the above-described (r, X, T) triplets with $X_i = y_i$. Allowing the volume expansion factor (which because of dilution is almost 1 in this case) to disappear into a unified frequency factor A_r, the six parameters that need to be optimized are: A_r, A_{CO}, A_{O2}, E_r, E_{CO} and E_{O2}. One could gain additional flexibility for fitting by making m and n available as free parameters but this would make the rate expression empirical rather than mechanistic. Here, then, is an example of judgment over objectivity.

Despite the large number of (r, X, T) points used, a unique and stable choice of these six parameters is not a simple matter. Discarding the MAM model as inadequate was relatively easy since it was impossible to obtain a MAM fit that gave a sum of squares of residuals (SSR) less than a factor of 100 greater than several different fits of the DAM model. One could continue to search for a better fit of the MAM but at some point the decision to drop this model is likely to become justified, even to the most careful interpreter.

At this point the search for the best fit concentrates on parameter optimization. Where the difficulty was (and usually is) greatest was in the inability of available parameter optimization methods to zero in automatically on a unique set of parameters for the DAM model itself. After an exhaustive search of parameter space using experimental data from several experiments, a number of sets of parameters could be found to fit the kinetic data satisfactorily as far as the conventional criterion of fit, the SSR, is concerned. Table 11.1 shows two sets of acceptable parameters.

For set 1, using the more than 12,000 triplets made available in the four experiments, the sum of squares of residuals was SSR = $2.97\text{x}10^{-1}$. For set 2 the corresponding value is 2.12. This is a pretty good fit for most purposes but not as good as set 1. The differences in parameter values are generally small except for A_{02}. However, the commercial programs used to optimize the fitted parameters were unable to arrive at set 1 from set 2. The procedure adopted, therefore, involved an extensive manual search designed to arrive at the final set of parameters reported in Table 11.1 as set 1. Unfortunately there is no guarantee that such a manual search will be successful at arriving at the best available fit, not to mention the "correct" set of parameters. The judgment called for in Chapter 7, Chapter 10, and elsewhere in the above text may lead one to chose a set of parameters that yields a greater-than-minimum sum of squares of deviations. The defense of the resultant choice of parameters is up to the reporter.

Table 11.1 *Parameter estimates for the DAM model of CO oxidation.*

Parameter	Set 1	Set 2
A_r [atm s^{-1}]	1.443×10^{16}	8.628×10^{15}
ΔE_r [J mol^{-1}]	1.462×10^{5}	1.444×10^{5}
A_{CO} [atm]$^{-1}$	6.832×10^{1}	5.306×10^{-1}
ΔH_{CO} [J mol^{-1}]	-7.495×10^{3}	-3.713×10^{4}
A_{O2} [atm]$^{-1}$	1.991×10^{-6}	1.053×10^{-11}
ΔH_{O2} [J mol^{-1}]	-8.299×10^{4}	-1.534×10^{5}

Part of the reason for this uncertainty is the region of parameter space investigated in this particular study. The final set of parameters reveals that the rate surface is by no means as simple as Figure 11.5 suggests. Instead, the extended isothermal reaction surface, when plotted on the composition plane, shows a maximum in rate, as shown by the contours on the specialized presentation of the rate surface shown in Figure 11.6. The contours show that a maximum rate exists at low concentrations of CO combined with appropriate concentrations of oxygen and that steep gradients are encountered in the vicinity of the maximum. This is the region containing an important feature of the rate surface and is where data *should* be collected in order to best define the rate expression.

Such low concentrations of CO usually occur near the end of a reaction and are not normally used as starting compositions. The sloping line shows what would be the composition trajectory of one of the starting compositions used in this study, if it were allowed to react in an isothermal PFR. The composition would drift downward and to the left as reactant was consumed, leaving some oxygen unreacted after all the CO had been consumed. Looking at the behaviour of the rate as the composition changed along this trajectory, one would observe that, up to a high level of conversion, the rate increases with conversion. Figure 11.7 shows this behaviour on another set of coordinates. Unfortunately, it is precisely in the region of highest rates, as conversion heads toward 100%, that temperature is also highest in a typical TS-PFR experiment, due to ramping and exothermicity, and few analyses can be captured before auto-ignition takes place.

Nevertheless, the existence of a maximum rate suggests that the reaction should be investigated in the region of this maximum in order to observe the most pronounced curvature on the kinetic surface and thereby get a more certain fit of all the parameters. Unfortunately the location, or even the existence, of the maximum is not known before the kinetic work was done. As a result, the data collected does not fall in the region of this maximum except by chance. A follow-up study should have been done in this case to examine the region of the predicted maximum, but in seeking to collect such data we run into an unexpected problem: it is not immediately clear what temperature scanning conditions will best delineate the region of this maximum. We find out this information only after fitting the experimental data to a rate expression and then using the expression obtained to simulate the behaviour of the reaction.

The difficulty of foreseeing the behaviour of a system at familiar isothermal conditions from TS-PFR data, or vice versa, of planning experiments such as those required above, is not yet well in hand. Procedures for dealing with this problem will eventually have to involve graphical presentation of superimposed isothermal plots, such as that on Figure 11.6 or its corresponding 3D form. Plots of this kind will allow the user to visualize the time course of events during a single ramp. This will help in planning follow-up experiments, since simulation of TS-PFR operation to predict the conditions required to achieve specified isothermal conditions is unlikely to be possible.

This problem arises because we have no means of assessing the actual thermal regime of the reactor during the experiment. It appears that only progress in simulation, data presentation and new methods of visualizing the behaviour of the system will make it possible to make full use of the information available from preliminary temperature scanning studies to guide subsequent experiments. In practice it will obviously be necessary to process data as soon as it becomes available and to examine the results in various coordinates including simulated presentations in isothermal conditions. Even now these efforts can serve to guide subsequent experiments so that a thorough investigation can be undertaken in the vicinity of "interesting" features on the kinetic surface before the reactor is assigned to another application.

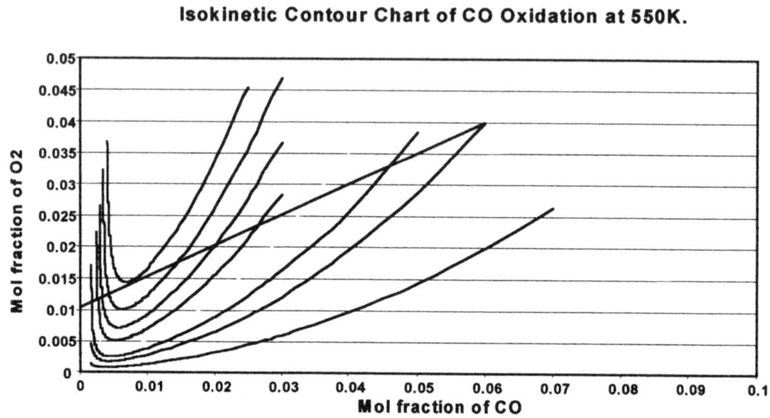

Figure 11.6[7]
Isokinetic contours at 550K and 1 bar. The lowest rate contour is to the right. It is evident that the contours form a ridge at low carbon monoxide mol fractions. The isothermal reaction path of a feed composition with CO/O_2 ratio of 0.06/0.04 is shown overlaid on the rate contours as a straight line sloping down to the left. As reaction progresses, the reactant CO/O_2 ratio decreases and reaction rate increases. After the path crosses the ridge of maximum rates, at a high conversion, the rate begins to decrease. A different mapping of this path is shown in Figure 11.7.

Considerations Regarding the Planning of TS-PFR Experiments

The data that allows the best opportunity for model discrimination in CO oxidation is that in the region of 70% to 95% conversion, using the starting compositions reported here. In the reported study of CO oxidation, data in this region was sparse, and it is rarely available in general. At these high conversions the feed tends to ignite, making the reaction no longer catalytic. Specially designed runs, in which auto ignition is avoided at conditions where the required composition is present, must be undertaken in follow-up work once the location of such interesting regions is identified by the preliminary studies. It may even be necessary to modify the physical configuration of the reactor/oven setup to prevent ignition or to allow low concentration feeds to be prepared, and thereby access the region of interest. Fortunately the oven/reactor unit is a stand-alone item in the TSR configuration (see Chapter 13) and can be readily modified or replaced, and mass flow meters covering various flow ranges are available. The advanced capabilities of the TS-PFR do not eliminate the need for insightful experimentation or ingenious equipment design.

---

[7] Reprinted from Applied Catalysis, Vol.190A, B.W. Wojciechowski et al., "Kinetic studies using temperature scanning: the oxidation of carbon monoxide," pp. 1-24 (2000) with permission from Elsevier Science.

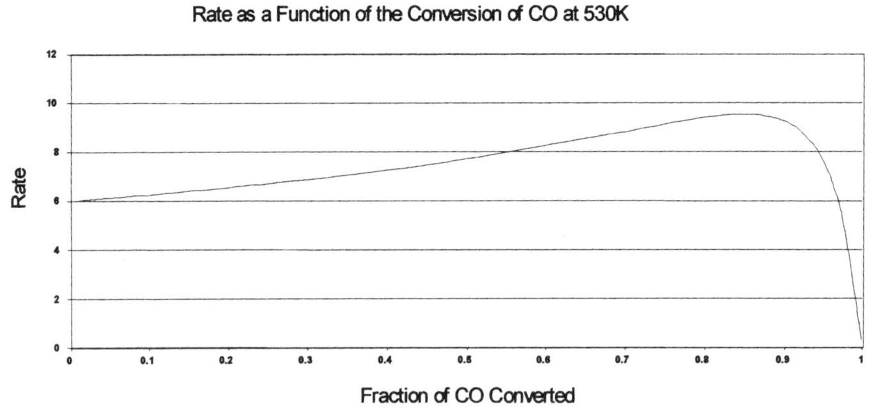

Figure 11.7[8]
Rate trajectories along reaction paths such as that shown by the straight line in Figure 11.6 show a maximum in rate at high conversion. This does not constitute auto-catalysis but results from the form of the rate expression, the relative sizes of the rate parameters, and the feed composition used. The phenomenon is ephemeral and will appear under some reaction conditions and not others, causing bewilderment and leaving much room for disputes between studies from different laboratories.

It is worthwhile to restate once more why such iterations between parameter fitting and experimentation are necessary, even in the presence of massive data collection made possible by the TSR. Consider Figure 11.5. If data is available only from the region of 0 to 70% conversion, the surface being fitted has little curvature, as is the case with the data used here, and could well be fitted by numerous forms of rate expression. On the other hand, in the region between 70% and 95% conversion the curvature is strong if the reaction mechanism is the DAM, and an interesting feature appears as a maximum rate at a specific reactant composition. The existence of this feature, if it is confirmed, severely limits the form of the applicable rate expression. If this behaviour is confirmed over the broad range of conditions that a TS-PFR data set makes available, the chances that a successful identification of a unique, perhaps "correct", rate expression are significantly increased. However, since one does not know *a priori* where or if such a feature exists, it is wise to plan kinetic investigations to proceed in several steps.

The initial kinetic study is intended to identify the reaction conditions of interest, if they are not already known from the literature, and uses a broad selection of operating conditions in a set of trial temperature rampings. From the conversions and other analytical information made available by this search, a set of temperatures, feed compositions, pressures and space times is selected. These are used to program a proper TSR experiment at one composition and pressure. Several experiments at various feed compositions need to be done subsequently.

[8] Reprinted from Applied Catalysis, Vol.190A, B.W. Wojciechowski et al., "Kinetic studies using temperature scanning: the oxidation of carbon monoxide," pp. 1-24 (2000) with permission from Elsevier Science.

The data from the first TSR experiment is used to search for the most likely form of the rate expression, as was done in this case. Often the candidate rate expressions are the result of mechanistic considerations, in which case understanding of the mechanism can serve as a guide to indicate fruitful ways of varying the reaction conditions in subsequent experiments.

The rate expression that results from the preliminary fit is then investigated in some detail by computer simulation, searching for conditions that will reveal its unique characteristics, say in the way outlined in Figures 11.6 and 11.7. The definitive experiments are then designed and carried out at conditions that allow for the best model discrimination. This was not done in the CO study, as the authors were principally interested in defining and verifying TS-PFR procedures rather than in investigating a specific kinetic mechanism.

Examining the Behaviour of Rate Expressions

The mechanism of a catalytic reaction is not likely to change with changes in catalyst composition or method of preparation, as long as the chemical nature of the catalyst is not changed. After all, a catalytic reaction mechanism involves chemisorption of the reactants and/or their fragments on active sites that depend largely on the composition of the catalyst. This means that the mechanism and therefore the overall rate expression can be expected to remain the same from sample to sample in a homologous series of catalysts.

At the same time, the absolute rate and selectivity of individual samples of catalyst from a homologous series depend on the relative rates of intervening, or chain propagating, reactions in the mechanism. As a result, the selectivity and overall parameters of the rate expression will be different from sample to sample. It is the unstated aim of most catalyst development efforts to find a catalyst formulation *whose kinetic parameters* give the desired selectivity and activity. Unfortunately this is not a common understanding of the process of catalyst development. Instead, thinking is focused on the chemical or physical aspects of the formulation and on the results of a standard-test evaluation of performance.

> *In a rational program of development of a new or improved catalyst it is essential to establish the mechanistic, or even an empirical but widely applicable, rate expression for the process, early in the development program.*

Although an empirical rate expression will fulfill some of the requirements, only a mechanistic expression can explain the effects of changes in catalyst formulation in terms of their influence on the elementary aspects of the reaction. Test results in catalyst development programs are rarely interpreted in terms of the kinetic parameters that determine the observed behaviour.

Once a believable and well-fitted mechanistic rate expression has been identified it is possible to search the parameter space of that equation, looking for improvements, their scale, and any interesting possibilities that may be presented by specific parameter sets. For example, a cursory search of the parameter space of the DAM model reveals the unexpected behaviour already shown in Figure 9.2. The same rate expression as that used to fit the experimental data from the TS-PFR study yields this bizarre result, where the apparent activation energy (or temperature coefficient) of the catalytic reaction is positive over some range of temperatures and then becomes negative. Figure 9.2 shows the behaviour of the rate as a function of temperature when the parameters, instead of being the ones reported in Table 11.1, are the alternate set cited in Chapter 9.

The literature contains hints of such negative activation energy observations, encouraging one in the belief that certain Pt/alumina formulations may in fact show this, and perhaps other, unexpected behaviour. Imagine how much confusion can be introduced into research in this topic alone by the presentation of two reports describing these two very different types of behaviour on apparently similar catalysts. Yet the mechanism and the rate equation for both types of behaviour are the same and the radically different behaviours obey the same analytical form of rate expression. The only difference between "normal" and "abnormal" behaviour is the size of the rate parameters.

Interestingly enough, this negative set of parameters fits one of the TS-PFR experiments from the cited work quite well. The data at a reactant composition where $CO/O_2 = 0.06/0.04$ is well fitted, as judged by the SSR, by the "abnormal" parameters of Chapter 9. At the same time, it was impossible to fit this set of parameters to all four TS-PFR experiments with anywhere near the sum of squares of residuals obtained using set 1. Even for the $CO/O_2 = 0.06/0.04$ data alone, a plot of error vs. rate clearly shows a systematic deviation of the calculated values from the experiment. Such distribution-of-error plots are seldom reported in the literature. They should be, but the available data from a testing program, or even from a more fundamental study, is rarely adequate for this kind of consideration. Nonetheless, it must be understood that:

> *the sum of squares of residuals (SSR) alone is an inadequate basis for accepting a best-fit set of parameters.*

The value of a mechanistic understanding of the rate of reaction

A deeper understanding of reaction mechanisms in catalysis is highly recommended to practitioners in this field, despite the difficulty of identifying a unique rate expression for many, if not all, bimolecular catalytic reactions, as illustrated in the rather simple reaction considered here. In particular, the difficulty of mechanism identification is overwhelming if just one feed composition has been studied. Even when one has available the wealth of data from a TS-PFR experiment, the best fit is hard to find on the basis of one experiment. Regardless of this complication, the rewards of understanding the details of the reaction mechanism and its kinetics are substantial.

It is obvious why a great deal of uncertainty hovers over kinetic interpretations made on the basis of the sparse and error-prone data obtained using conventional methods of experimentation. This uncertainty is sometimes compounded by confusion brought about by incongruous results reported by different laboratories. Such disagreements can arise from various causes, but most often because of a lack of consistency in defining the space time and its units and/or other influential aspects of reaction conditions or reactor configuration. On top of this, apparent incongruities may be rooted in systematic error, poor data handling procedures, questionable interpretation of observations, or differences in the catalyst itself, rather than being the result of erroneous readings of reactor outputs per se.

There is reason to believe that catalyst samples prepared in different laboratories may not be identical, despite the best efforts to make them so. The most likely differences between samples are probably due to differences in adsorption properties, which are rarely well identified during catalyst characterization. Yet differences in adsorption equilibrium constants can lead to great variability in kinetic and selectivity behaviours, even in simple catalytic rate expressions, as we have just seen.

Looking at a mechanistic catalytic rate expression, it can be seen that changes in the rate constant parameter k_r alone (and attention is focused on this in most catalyst development programs) generally cause simple alterations in the observable rate and selectivity phenomena. In fact, the procedure routinely used in testing catalysts at standard conditions presumes that successive samples of development catalysts will differ *in this parameter alone*. This would usually result in simply shifting the (r, X, T) rate surface up or down, (see Figure 7.8 in Chapter 7 for an example of a broadly assessed surface) changing its vertical (rate axis) *position* without changing its shape. In that case the *standard run*, which measures one point on this surface, will correctly reveal how much change to expect at other points on the reaction rate surface.

Changes in the adsorption parameters, on the other hand, always lead to changes in both the position and the shape of the surface. In that case the traditional standard run provides information *only* at the standard experimental condition; it is more than likely to be misleading if the result is used to interpret behaviour elsewhere on the surface.

There is a further question worth examining. Is it likely that one can change the rate constant of a catalytic rate expression without changing some or all of the adsorption constants? After all, the catalytic rate constant involves nothing unusual from the point of view of kinetics. It involves one of a set of surface-dependent reactions taking part in the overall mechanism. If that is so, other reactions in the set are likely to be affected any time we expect to see a change in the rate constant. It is then reasonable to expect that the shape of the kinetic surface will change with *all* changes in activity. However, it is impossible to predict the resultant shape without a thorough understanding of the effects of changes in catalyst formulation on rate parameters and thence, by simulation using the rate expression, of the kinetic surface. In other words, it is always inherently risky to use the results of a standard test to estimate catalyst performance of any other homologous catalyst formulation in a development program. It is entirely possible that new catalyst formulations will exhibit enhanced activity or selectivity at other than the standard test conditions and that these enhancements will be missed by conventional routine testing.

> *A standard test in catalyst development work is justified only if one is committed to continue commercial operation at the conditions of the standard test, ignoring any benefits that may be available from operating at conditions that accommodate the properties of a new or improved catalyst.*

Steam Reforming of Methanol

The Mechanism

Steam reforming of methanol presents a much more intricate reaction mechanism and a considerably more complicated chemistry than the oxidation of carbon monoxide. The reaction can be formally separated into three highly coupled equilibria.

$$CH_3OH + H_2O \underset{-k_R}{\overset{k_R}{\rightleftarrows}} CO_2 + 3H_2$$

$$CO + H_2O \underset{-k_W}{\overset{k_W}{\rightleftarrows}} CO_2 + H_2$$

$$CH_3OH \underset{-k_D}{\overset{k_D}{\rightleftarrows}} CO + 2H_2$$

These concurrent equilibria are coupled not only via common reactants and products but also by sharing surface-resident intermediates that lie on the path of the overall rate of reaction.

Examination of the above equilibria shows that three products appear: H_2, CO_2 and CO. Each of these products will have its own mechanism and rate of formation. Each of these rates will be coupled to the rates of formation of the other products, but will not be directly proportional to them. A full overall mechanism for this reaction is thought (Peppley, B.A. et al. (1999)) to consist of over twenty elementary surface reactions taking place on two different types of sites. It is assumed in the development of this mechanism that the two types of sites are energetically homogeneous within each type.

Steady state solutions have been invoked in developing rate expressions from this mechanism for each of the observed products. This allows the use of mechanistic kinetic rate expressions for the formation of the products rather than having to deal with the full twenty-reaction network of elementary reactions. A twenty-reaction network would require at least forty parameters to be fitted in correlating data from this system. As we will, see this number is considerably reduced by solving the mechanism at steady state and fitting the resultant steady state rate expressions.

The Rate Expressions

The mechanistic rate expressions for the appearance of the three products are:

Methanol-Steam Reaction

$$r_R = \frac{k_R K^*_{CH_3O'} \dfrac{p_{CH_3OH}}{p_{H_2}^{1/2}} \left(1 - \dfrac{p_{H_2}^3 p_{CO_2}}{K_R p_{CH_3OH} p_{H_2O}}\right) C^T_{S_1} C^T_{S_{1a}}}{\left(1 + K^*_{CH_3O'} \dfrac{p_{CH_3OH}}{p_{H_2}^{1/2}} + K^*_{HCOO'} p_{CO_2} p_{H_2}^{1/2} + K^*_{OH'} \dfrac{p_{H_2O}}{p_{H_2}^{1/2}}\right)\left(1 + K_{H_{1a}}^{1/2} p_{H_2}^{1/2}\right)} \tag{11.6}$$

Water-Gas Shift Reaction

$$r_W = \frac{k^*_W K^*_{OH'} \dfrac{p_{CO} p_{H_2O}}{p_{H_2}^{1/2}} \left(1 - \dfrac{p_{H_2} p_{CO_2}}{K_W p_{CO} p_{H_2O}}\right)\left(C^T_{S_1}\right)^2}{\left(1 + K^*_{CH_3O'} \dfrac{p_{CH_3OH}}{p_{H_2}^{1/2}} + K^*_{HCOO'} p_{CO_2} p_{H_2}^{1/2} + K^*_{OH'} \dfrac{p_{H_2O}}{p_{H_2}^{1/2}}\right)^2} \tag{11.7}$$

Decomposition Reaction

$$r_D = \frac{k_D K^*_{CH_3O''} \dfrac{p_{CH_3OH}}{p_{H_2}^{1/2}} \left(1 - \dfrac{p_{H_2}^2 p_{CO}}{K_D p_{CH_3OH}}\right) C^T_{S_2} C^T_{S_{2a}}}{\left(1 + K^*_{CH_3O''} \dfrac{p_{CH_3OH}}{p_{H_2}^{1/2}} + K^*_{OH''} \dfrac{p_{H_2O}}{p_{H_2}^{1/2}}\right)\left(1 + K_{H_{2a}}^{1/2} p_{H_2}^{1/2}\right)} \tag{11.8}$$

where:

$$k_R = k_R^\infty \exp\left(\frac{-E_R}{RT}\right) \tag{11.9}$$

$$k_D = k_D^\infty \exp\left(\frac{-E_D}{RT}\right) \tag{11.10}$$

$$k^*_W = k_W^{\infty *} \exp\left(\frac{-E^*_W}{RT}\right) \tag{11.11}$$

$$K^{*}_{CH_3O'} = \exp\left(\frac{\Delta S^{*}_{CH_3O'}}{R} - \frac{\Delta H^{*}_{CH_3O'}}{RT}\right) \tag{11.12}$$

$$K^{*}_{HCOO'} = \exp\left(\frac{\Delta S^{*}_{HCOO'}}{R} - \frac{\Delta H^{*}_{HCOO'}}{RT}\right) \tag{11.13}$$

$$K^{*}_{OH'} = \exp\left(\frac{\Delta S^{*}_{OH'}}{R} - \frac{\Delta H^{*}_{OH'}}{RT}\right) \tag{11.14}$$

$$K_{CO_2'} = \exp\left(\frac{\Delta S_{CO_2'}}{R} - \frac{\Delta H_{CO_2'}}{RT}\right) \tag{11.15}$$

$$K_{H(1a)} = \exp\left(\frac{\Delta S_{H_{1a}}}{R} - \frac{\Delta H_{H_{1a}}}{RT}\right) \tag{11.16}$$

$$K^{*}_{CH_3O''} = \exp\left(\frac{\Delta S^{*}_{CH_3O''}}{R} - \frac{\Delta H^{*}_{CH_3O''}}{RT}\right) \tag{11.17}$$

$$K^{*}_{HCOO'} = \exp\left(\frac{\Delta S^{*}_{HCOO'}}{R} - \frac{\Delta H^{*}_{HCOO'}}{RT}\right) \tag{11.18}$$

$$K^{*}_{OH''} = \exp\left(\frac{\Delta S^{*}_{OH''}}{R} - \frac{\Delta H^{*}_{OH''}}{RT}\right) \tag{11.19}$$

$$K_{CO_2''} = \exp\left(\frac{\Delta S_{CO_2''}}{R} - \frac{\Delta H_{CO_2''}}{RT}\right) \tag{11.20}$$

$$K_{H_{2a}} = \exp\left(\frac{\Delta S_{H_{2a}}}{R} - \frac{\Delta H_{H_{2a}}}{RT}\right) \tag{11.21}$$

The values of surface site concentrations used were reported as:

$$C^{T}_{S_1} = 7.5E-06 \qquad C^{T}_{S_{1a}} = 1.5E-05$$
$$C^{T}_{S_2} = 7.5E-06 \qquad C^{T}_{2a} = 1.5E-05$$

all in (mol m^{-2}) of catalyst surface.

The surface per gram was reported as 101.72 (m^2 $g.cat^{-1}$). The magnitudes of the surface concentrations of sites influence only the size of the frequency factors of the rate constants and might just as well have been left unspecified. Unless these quantities are obtained from theoretical considerations or are measured in separate experiments, their values have little meaning. Here they are included in the rate constants.

Due to the application of steady state, the mechanistic overall rate equations are fairly compact and the number of parameters to be fitted has been cut by half, to twenty. This is still a large number of parameters and much data is required to obtain a reliable fit. The main reason for the reduction in the number of parameters to be evaluated is that the parameters marked by an asterisk contain quotients of several more fundamental constants. Nevertheless, in arriving at these rate expressions, no important mechanistic information is left out. If the mechanism had been treated as a network of reactions, all forty parameters would have had to be evaluated individually. As long as steady state is achieved rapidly, there is significant benefit in developing mechanistic rate expressions rather than dealing with full reaction networks.

Data Fitting

Despite the simplification afforded by the steady state assumption, the fitting of the experimental data in this case requires the optimization of no less than twenty parameters. The effort required to gather adequate data by conventional methods of experimentation makes a kinetic study of this system less than attractive. Fortunately, a TSR can easily provide the rate measurements required for a reliable fitting of a mechanism of this complexity.

The fitting procedures themselves are further complicated by the fact that each of the products is involved in more than one of the three equilibria listed above. We can see from the equilibria that, in terms of the individual reaction rates presented above, the rates of formation of the products are:

$$r_{CO_2} = (r_R + r_W) \tag{11.22}$$

$$r_{CO} = (r_D - r_W) \tag{11.23}$$

$$r_{H_2} = (3r_R + 2r_D + r_W) \tag{11.24}$$

while the rates of consumption of the reactants involve:

$$-r_{H_2O} = (r_R + r_W) \tag{11.25}$$

$$-r_{CH_3OH} = (r_R + r_D) \tag{11.26}$$

Because of the definition of space time, the rates of conversion to CO and CO_2 are reported in (mol $g.cat^{-1}$ sec^{-1}). The conversion from partial pressures in the mechanistic rate expressions to mol fractions and hence to fractional conversions that finally result in these units was done in the way reported for the case of carbon monoxide oxidation. This leaves factors such as $\delta/Py_{Me,in}$ on the right-hand side, and affects the units of the rate constants. In this case several pressures were studied so that this factor was not incorporated in the frequency factors of the rate constants. It was explicitly evaluated at each condition and the rate constant multiplied by the result.

The various mass and atomic balance considerations presented in Chapter 7 offer useful constraints on the rates in this system and others where multiple products appear. In principle the rates themselves, both in the form of experimental results and as calculated values from the final rate expressions, should be in mass and atomic balance over their entire range of applicability.

TS-PFR Results in Methanol Reforming

The raw data for steam reforming of methanol was collected at 0.75 steam/methanol ratio, using 30 mol % dilution with argon, at a total reactor pressure of 50 psig. The collected data was converted to fractional conversions and yields, and was filtered. In this case results for the production of CO and CO_2 as well as methanol consumption were the principal results involved in parameter estimation, requiring that all their rate surfaces be smoothed by filtering. Not only are these products easier to analyze than hydrogen and water, but they are sufficient to define all the required rate parameters. A more elaborate study might involve fitting the other components independently in order to confirm the validity of parameters obtained. Balancing of product analyses to optimize mass and atomic balances and produce reactant yield surfaces subject to these balance constraints was not done.

Figure 11.8 shows a 2D plot of the mole fraction of methanol converted before (jagged trace) and after (smooth curve) 1-dimensional filtering. This type of filtering is satisfactory for removing noise in one dimension, along the curve, but leaves untouched the noise in the other important dimension, the separation between curves. This results in corrugations in the surface of (X, τ, T). Figure 11.5 showed such corrugations in the previous example dealing with carbon monoxide oxidation. This will not do, where slopes must be taken off this surface in order to calculate the rates $r = dX/d\tau$.

To remove this noise one must perform a 2-dimensional filtering of the data, both along the curves and across the curves, and thereby obtain the smooth (X, τ, T) surface required for rate extraction. The filtering can be carried out on the raw data or in two steps, the 2D filtering being done after the initial 1D filtering shown in Figure 11.8. It is not immediately clear which is the better procedure to use.

From the 2D-filtered data for CO_2 and CO production, smooth rate surfaces are generated. From these, as in the case of CO oxidation, sets of (r, X, T) triplets required for data fitting to the candidate rate expressions were sieved out. First several isothermal sets of sieved (r, X, $T_{e,const}$) data are fitted to isothermal forms of the rate expression and the constants obtained plotted on Arrhenius plots. The Arrhenius parameters from these plots were then introduced as starting values for an "all-up" fit of the rate expression. The parameters were then optimized for that set of starting values. Other (similar) sets of starting values were then tested to see if a better optimum fit could be obtained or if all optimizations converged at the same optimum. Such a fitting of the CO_2 and CO production rates resulted in the parameters shown in Table 11.2.

Examination of the parameters shows that the energy terms in Table 11.2 fall within the limits delineated in the discussion above, as did those in Table 11.1. As in Table 11.1, some of the heats of adsorption in Table 11.2 turn out to be endothermic. This should not come as a surprise (see also Chapter 9) since some of the adsorbed species in the mechanism are fragments of the reactants and products. The chemistry which lies behind this observation invites further thought and verification.

The units reported here differ from those used in the oxidation of carbon monoxide in that bar is used in place of atmosphere for pressure measurements and tau was defined in terms of grams of catalyst divided by the molar flow rate of the feed, resulting in the units of rate being (mol $g.cat^{-1}$ $s^{-1}l$).

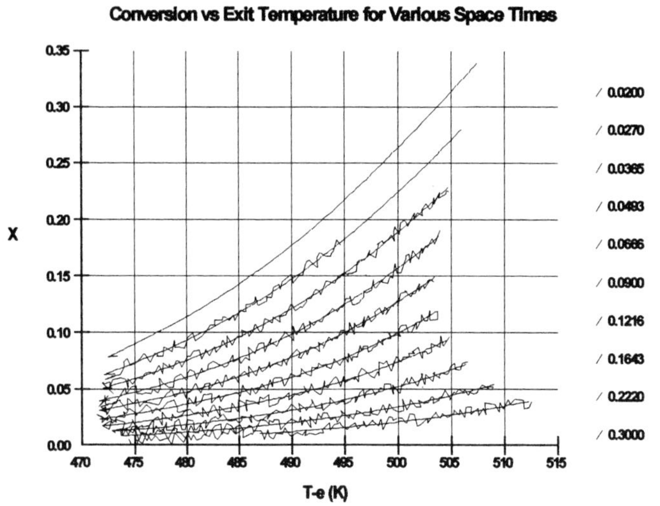

Figure 11.8[9]

Raw and filtered experimental data for the reforming of methanol. The jagged trace in each case is the result of calculating instantaneous conversion at each analysis. The smooth curves underlying the experimental data are the result of applying a 1D filter to data from each run in turn. This leaves unsmoothed any errors between curves. Such errors can arise from flow controller inaccuracies, temperature variations and other causes and will lead to a bumpy conversion surface if left untreated.

One of the inconveniences of TS-PFR methods is that it is difficult, in fact impossible, to compare conversions calculated using the fitted rate expression with raw TS-PFR data. This point was raised previously in connection with the investigation of carbon monoxide oxidation. One can simulate kinetic behaviour using the above equations and parameters to produce the expected *isothermal* behaviour of the system but not that observed in the experimental results that are obtained from the TS-PFR during temperature ramping. This is unavoidable and results from our lack of knowledge of the axial temperature profile in the experimental set-up.

[9] Reprinted from Applied Catalysis, Vol 179A, B.W. Wojciechowski et al., "Kinetic studies using temperature scanning: the steam reforming of methanol", pp. 51-70 (1999) with permission from Elsevier Science.

Table 11.2

Parameters for the Kinetic Model of Steam Reforming of Methanol on BASF K3-110, A Cu/ZnO/Al2O3 Catalyst.

Rate Constant or Equilibrium Constant		ΔS (J mol^{-1} K^{-1}) or k^∞ (bar mol (g.cat)$^{-1}$s^{-1})	ΔE or ΔH (kJ mol^{-1})
k_R	**(bar mol (g.cat)$^{-1}$s^{-1})**	**6.4E+14**	**101.8**
$K^*_{CH3O'}$	(bar$^{-0.5}$)	51.7	-20.9
$K^*_{OH'}$	(bar$^{-0.5}$)	-27.3	-7.1
$K_{H(1a)}$	(bar^{-1})	-86.0	-31.9
K^*_{HCOO}	(bar$^{-1.5}$)	75.8	40.5
k_D	**(bar mol (g.cat)$^{-1}$s^{-1})**	**1.4E+11**	**70.0**
$K^*_{CH3O''}$	(bar$^{-0.5}$)	-48.7	-30.6
$K^*_{OH''}$	(bar$^{-0.5}$)	25.6	-9.7
$K_{H(2a)}$	(bar^{-1})	-22.6	-23.7
k^*_W	**(bar mol (g.cat)$^{-1}$s^{-1})**	**7.0E+13**	**81.0**
K_W	(bar^0)	-2.029	17.2
K_D	(bar^2)	12.621	-42.7
K_R	$= K_D * K_W$		

Uses of the Parity Plot

The preferred way of checking the validity of the fitted rate parameters in TSR work (and in general, in cases when the data ranges over a wide selection of reaction conditions) is to construct *parity plots* (see Chapter 10). In TSR work this is done using experimental rates from the TSR, calculated from the slopes on filtered raw data surfaces, and rates calculated from the rate expression fitted to these rates. A plot of one against the other reveals how well the fitted rate expression conforms to the rate surface obtained from raw experimental results. The parity plots offer a 2-dimensional comparison of all the experimental rate data gathered in any number of TS-PFR experiments, against the corresponding calculated rates of reaction, regardless of temperature, pressure, or feed composition, in each experiment.

Although the parity plots do not allow a direct comparison of calculated results with experimental *readings,* they do allow us to examine the fit between predicted rates and those extracted from the TS-PFR experiment and used in the parameter optimization. Deviant behaviour is evidenced on parity plots as excessive scatter from the diagonal or by trends in the scatter. Unsatisfactory correlations will exhibit well-defined trends on parity plots, even if the SSR for the whole data set is deemed to be acceptable. Such trends can then be used to identify the shortcomings of the proposed rate expression.

Two parity plots for the data on the steam reforming of methanol are shown on Figures 11.9 and 11.10. Since two rates of product formation are extracted from the steam reforming data, for CO_2 and CO respectively, we need to examine the parity of both rates with predictions. We see at once that the data for CO (Figure 10.10) shows more scatter than does the data for CO_2. This is probably due to the much smaller amounts of CO in the products and hence a less accurate analysis for CO. This difference survives the filtering procedure, which, in order to avoid distortions that arise with "over-filtering", is carried out so as to leave some noise (roughness) on the kinetic surface, and hence some error in the calculated rates.

The residual noise is small but the danger of over-filtering the data and distorting the shape of the rate surface is ever-present and must be considered and avoided in the early stages of data treatment. How much residual noise to leave in is a matter of judgment that can be confirmed in various semi-quantitative ways. The recommended method of monitoring the filtering process is to generate a parity plot of the original raw data vs. the filtered data. Scatter may be large due to the pre-existing noise but systematic deviations from the 45° parity line serve as a warning that the underlying shape of the surface is being altered by the filtering process. When and if such deviations can be accepted is up to the interpreter.

A more demanding and more convincing test of the veracity of the parameters lies in calculating the rates of methanol conversion and/or product formation and comparing these to rates extracted from a new, as yet un-interpreted, TS-PFR experiment. The new experiment should be done under conditions where the available rate expression applies but whose data was not used in arriving at the currently available rate parameters. To confirm the generality of the parameters in Table 11.2 further testing of the model was done by running the reaction at several other pressures. When the same parameters as those in Table 11.2 were used, the parity plots for the rate of CO production, the most sensitive and industrially significant measurement, came out as shown in Figures 11.12, 11.13 and 11.14. Data in those plots came from experiments at 0, 40 and 60 psig operating pressure and can be compared to that shown in Figure 11.10. In all cases regression of the data yields a line very close to the 45° line starting at the origin.

This complex and interesting reaction yielded more data in the several TS-PFR runs performed than one could reasonably expect from a conventional isothermal study lasting many times longer. The data from the TS-PFR is cohesive and consistent, much more so than data collected one point at a time in a conventional setup. In addition, the application of sophisticated filtering techniques allowed the removal of much of the random noise present in the raw data. This in turn permitted a more certain identification of the appropriate rate expression and a better definition of its optimum parameters.

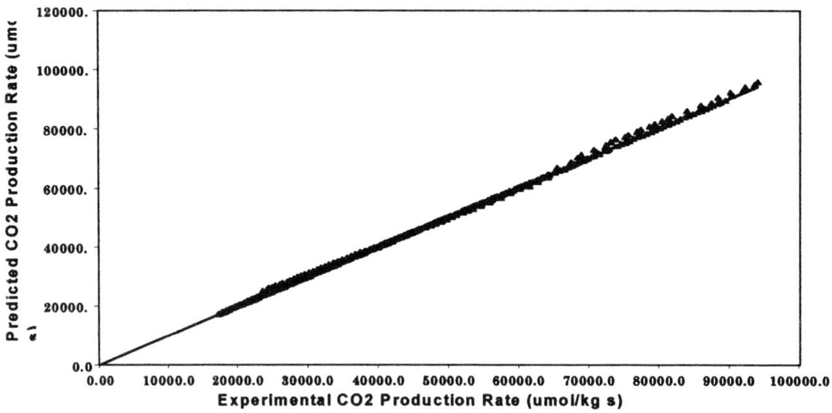

Figure 11.9[10]
Parity plot showing the predicted rate of CO_2 formation plotted against the experimentally obtained value. This presentation compares the predictions of the proposed rate expression to the results obtained, regardless of conversion, reactant composition or any other factors.

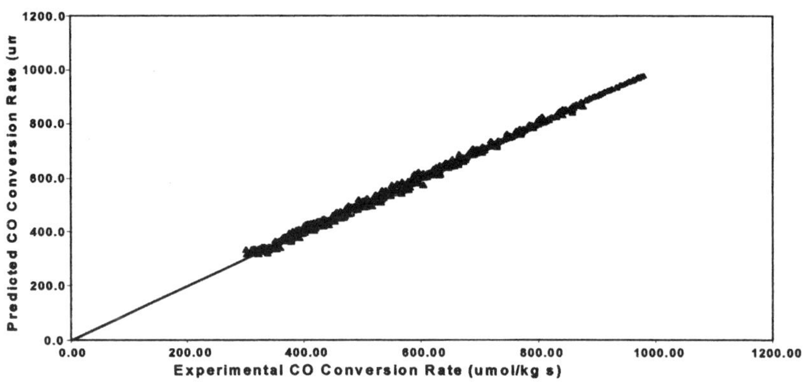

Figure 11.10[10]
Parity plot showing the predicted rate of CO formation plotted against the experimentally obtained value. This presentation compares the predictions of the proposed rate expression to the results obtained for an independently measured property of the reaction already considered in Figure 11.9. In principle such parity plots should be constructed for each component of the output stream.

[10] Reprinted from Applied Catalysis, Vol 179A, B.W. Wojciechowski et al., "Kinetic studies using temperature scanning: the steam reforming of methanol", pp. 51-70 (1999) with permission from Elsevier Science.

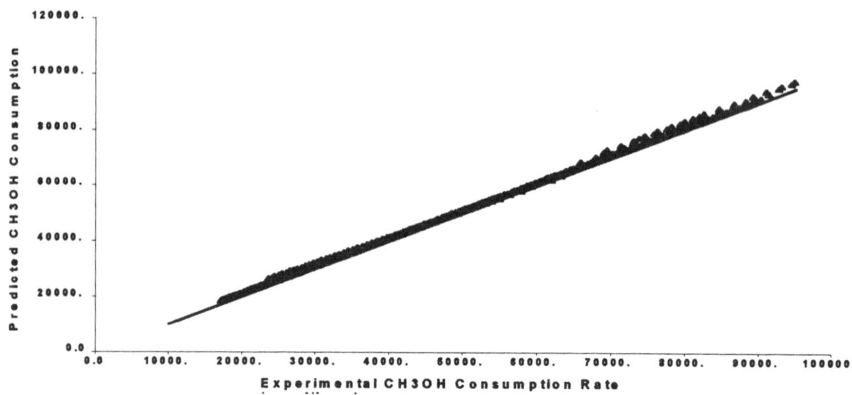

Figure 11.11[11]

Parity plot for methanol conversion in the base experiment. The potential significance of trends that one may or may not see in this Figure has to be judged. It was the authors' view that any discernible trends seen in Figure 11.11 are not significant.

General Observations

From the study described above it appears that the mechanism of reaction and its associated rate parameters do not change in the range of pressures and temperatures investigated. One can now say that:

1. the mechanism of this reaction is likely to be much like the one proposed by Peppley et al. (1999);
2. the kinetics of this reaction, on this catalyst, are now well established over a broad range of operating conditions;
3 changes in catalyst performance with incremental changes in catalyst formulation could now be observed as changes in the kinetic parameters of the same rate expressions;
4. correlation of parameter changes with changes in catalyst properties, combined with simulations of the behaviour of the established rate expressions, should reveal the rational route to an optimum catalyst formulation.
5. In case there is an interest in treating this reaction as a network, one could take the reactions composing the mechanism and proceed to solve them as a network. The parameters first obtained from the steady state mechanism could now serve as initial values or constraints on the elementary constants of the component reactions of the network.

---

[11] Reprinted from Applied Catalysis, Vol 179A, B.W. Wojciechowski et al., "Kinetic studies using temperature scanning: the steam reforming of methanol", pp. 51-70 (1999) with permission from Elsevier Science.

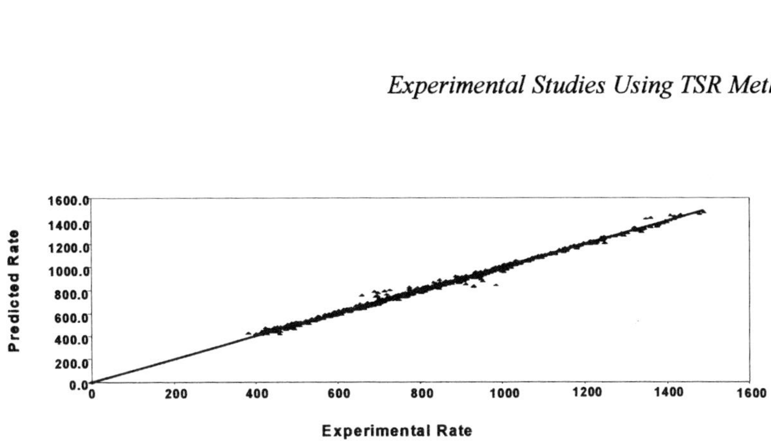

Figure 11.12[12]
Parity plot for CO formation at 0 psig reactor pressure

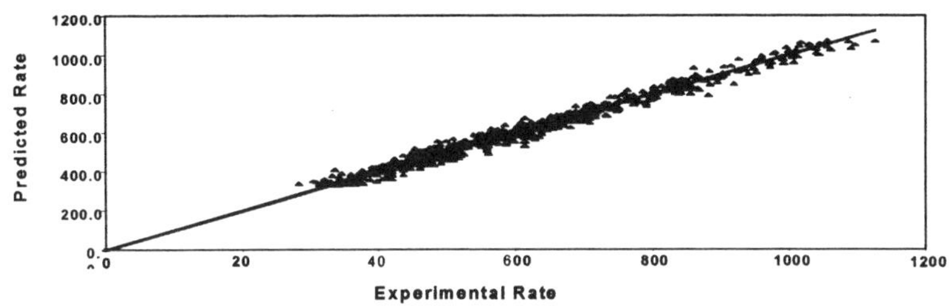

Figure 11.13[12]
Parity plot for CO formation at 40 psig reactor pressure.

Figure 11.14[12]
Parity plot for CO formation at 60 psig reactor pressure.

[12] Reprinted from Applied Catalysis, Vol 179A, B.W. Wojciechowski et al., "Kinetic studies using temperature scanning: the steam reforming of methanol", pp. 51-70 (1999) with permission from Elsevier Science.

The suggested simulations and the systematic examination of parameters vs. catalyst properties in point 4 have yet to be done. However, for the first time, such studies are now possible at reasonable cost.

> *Collections of rate parameters correlated with catalyst properties will offer a valuable resource in the ongoing attempts to make catalysis less of an art and more of a science. When this possibility is finally realized, kinetic studies will make the rational design of catalysts possible, and the pursuit of kinetic studies will have matured to a new level of utility.*

The Hydrolysis of Acetic Anhydride

The two previous examples dealt with gas phase catalytic reactions studied in a TS-PFR. Temperature scanning, however, is not limited to this type of reaction or reactor. It is a broadly applicable technique of experimentation, applicable to a variety of chemical reactions, in a variety of reactor types. It is rare, however, to find a reaction that can conveniently be carried out in a variety of reactors. One such reaction is the hydrolysis of acetic anhydride, a liquid phase reaction with particularly simple kinetics. This reaction can therefore be used to examine the consistency of data obtained from various reactors, as well as to provide an illustration of the application of the TSR technique to a homogeneous reaction in the liquid phase.

A second interesting property of the hydrolysis reaction is that it allows for the use of the integrated form of the rate expression first mentioned in Chapter 4. By now it is clear that the integral method of data interpretation is not particularly advantageous in TSR studies. Nevertheless, this procedure can be made to yield the one temperature dependent rate parameter of power-law rate expressions and, in the case of the TS-PFR and the TS-BR, it can yield this parameter in real time. A suitable data display can therefore be made to generate an Arrhenius plot of the first order rate constant as it is calculated after each analysis. The current regression estimate of the frequency factor and activation energy can then be displayed on the same plot. The system offers a fine way of doing kinetic studies in a number of similar aqueous-phase reactions, as well as offering a simple way of introducing temperature scanning methods to undergraduate laboratories.

The Chemistry of the Reaction

The overall hydrolysis of acetic anhydride reaction can be written:

$$CH_3-C(O)H-O-C(O)H-CH_3+H-O-H \rightarrow 2\,CH_3-C(O)OH$$

or, more succinctly,

$$Ac_2O+H_2O \rightarrow 2AcOH$$

The reaction mechanism is well documented (March, J., (1985)). This is not a classical SN substitution reaction but rather an addition reaction (Candruff, J., (1978)). The reaction mechanism proceeds via three irreversible steps:

1. addition,
2. elimination,
3. proton transfer to the solvent.

Step one, the addition reaction, is the rate controlling process.

The Kinetics of the Reaction

Since step one is rate controlling, the rate of the hydrolysis reaction can be written as:

$$\frac{-d[Ac_2O]}{dt} = k'[H_2O][Ac_2O] \tag{11.27}$$

However, under conditions where there is a large excess of water, pseudo-first-order reaction kinetics can be observed. The expression now simplifies to:

$$\frac{-d[Ac_2O]}{dt} = k[Ac_2O] \tag{11.28}$$

or,

$$\frac{-dC_A}{dt} = kC_A \tag{11.29}$$

where $k = k'\,[H_2O]$ is the observed rate constant (in units of time^{-1}) for the pseudo-first-order rate of hydrolysis. Note that the frequency factor of this constant contains a concentration term for water. This means that the absolute value of the pre-exponential of k will be anomalous, in terms of the issues raised in Chapter 9, as it represents $A = A'\,[H_2O]$ where A' is the true bimolecular frequency factor. By integrating this first order rate expression we obtain:

$$C_A = C_{A0}\exp(-kt) \tag{11.30}$$

where C_A denotes the concentration of acetic anhydride in solution. This is the simple yet mechanistic form of the integrated rate expression we will be examining using TSR data.

Measuring Conversion Using the Conductivity of the Solution

The hydrolysis of acetic anhydride was monitored using an electrode and a digital conductivity meter (Asprey, S.P. et al. (1996)). The general layout of the apparatus is shown in Figure 11.15.

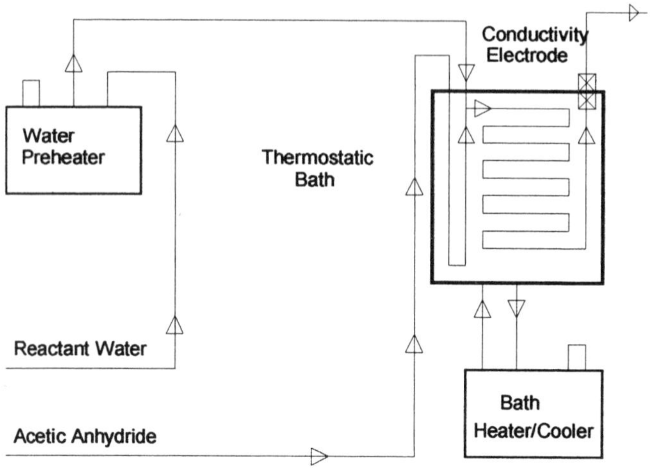

Figure 11.15
Schematic of a three-way liquid phase reactor setup. The setup can be operated as a TS-PFR in the configuration shown or converted to a TS-CSTR or a TS-BR by replacing the reactor coil in the thermostat with a stirred vessel.

Care had to be taken that the conductivity electrode was at the same temperature as the reactor bath and that the reagents were preheated to the bath temperature before being combined at the entrance to the reactor. Various other design considerations were taken into account, as they must be in each new reactor.

Acetic anhydride is a neutral molecule and therefore does not form ions. However, as the reaction progresses, ionic species are produced by the hydrolysis of the anhydride to form acetic acid, an ionizable molecule. As a result, the electrical conductivity of the solution will change with the concentration of the acetic acid formed (i.e. with the conversion of the anhydride), but not directly. The reason for this complication is that the acetic acid produced does not dissociate completely but forms an equilibrium with its ions:

$$AcOH \xleftrightarrow{K_{eq}} AcO^- + H^+$$

The equilibrium constant is defined by

$$K_{eq} = \frac{[AcO^-][H^+]}{[AcOH]} = \frac{[H^+]^2}{[AcOH]} \tag{11.31}$$

where it is taken that the concentration of hydrogen ions is the same as the concentration of acetate ions. The product ions on the right-hand side of the equilibrium are the conducting species. At low concentrations their capacity to carry charges, and hence the conductivity of the solution, varies linearly with the concentration of protons. We therefore write the total conductivity of the reacting solution as a linear function of proton concentration:

$$T_t = \lambda\,[H^+] + T_o \qquad (11.32)$$

where T_o = initial conductivity of the solution at time t = 0, when no protons beyond those in distilled water are present.

T_t = conductivity at time t during the course of the reaction when protons due to the ionization of the acetic acid produced are also present.

λ = unknown constant relating conductivity to the concentration of protons.

Using the expression for equilibrium, we can rewrite the conductivity relation in terms of the equilibrium constant and the concentration of acetic acid:

$$T_t = \lambda\sqrt{K_{eq}[AcOH]} + T_o \qquad (11.33)$$

Using this we can write the conductivity in terms of the acetic anhydride converted:

$$T_t = \lambda\sqrt{2K_{eq}\{[Ac_2O]_o - [Ac_2O]\}} + T_o \qquad (11.34)$$

When the reaction has gone to completion, the final conductivity will correspond to that resulting from the total conversion of the Ac_2O present in the original sample.

$$T_\infty = \lambda\sqrt{2K_{eq}[Ac_2O]_o} + T_o \qquad (11.35)$$

Formulating the Rate Expression in Terms of Conductivity

The above relationships can be assembled to give,

$$\frac{T_t - T_o}{T_\infty - T_o} = \sqrt{1 - \frac{[Ac_2O]}{[Ac_2O]_o}} = \sqrt{1 - \exp(-kt)} = \sqrt{1 - \frac{C_A}{C_{A0}}} \qquad (11.36)$$

where T_∞ denotes the conductivity at infinite time, when all the acetic anhydride has reacted to produce acetic acid,

We see that the integrated form of the kinetic expression containing the rate constant k is related to the concentration of reactants in the normal way, but those expressions are related indirectly to the measured conductivity T via the expression

$$\left(\frac{T_t - T_o}{T_\infty - To}\right)^2 = 1 - \frac{C_A}{C_{A0}} = 1 - \exp(-kt) \qquad (11.37)$$

and hence

$$\left[\frac{(T_\infty - T_0)^2 - (T_t - T_0)^2}{(T_\infty - T_0)^2}\right] = \frac{C_A}{C_{A0}} = \exp(-kt) \tag{11.38}$$

From the above we obtain an explicit expression for the rate constant in terms of measured quantities:

$$-k = \frac{1}{t}\ln\left[\frac{(T_\infty - T_0)^2 - (T_t - T_0)^2}{(T_\infty - T_0)^2}\right] \tag{11.39}$$

Since T_∞ and T_0 are pre-established values, and the form of the rate expression is known, each time the current conductivity T_t is measured one can calculate the current value of the rate constant k. This rate constant is different at each clock time during a TSR experiment because temperature is being ramped. By following the change in the rate constant with ramp temperature, in real time, on an Arrhenius plot of ln(k) vs. 1/T, the evolution of the best fit frequency factor and activation energy with increasing data availability and ramp temperature can be observed.

$$\ln(k) = \ln\left(\frac{-1}{t}\ln\left[\frac{(T_\infty - T_0)^2 - (T_t - T_0)^2}{(T_\infty - T_0)^2}\right]\right) = \ln(A) - (\Delta E_a / RT) \tag{11.40}$$

A plot of $\ln(f(T_i(t))$ vs. 1/T should give a straight line with slope $(-\Delta E_a/R)$, thus revealing the activation energy for the reaction. The pre-exponential factor is obtained from an extrapolation of this line to the y-intercept where the logarithm of the frequency factor, ln(A), is found at 1/T = 0.

Since conductivity in a given solution is a function of temperature, both T_0 and T_∞ must be known over the range of temperatures to be scanned in the given experiment. This allows the above function to be correctly evaluated at each temperature during the ramp. This information is obtained by ramping the temperature of suitable samples to obtain the values of each of T_0 and T_∞ over the temperature range of interest (a small range of 25°C to 55°C in this case). The resultant T_i data is correlated using a linear or some other suitable form of regression. Figure 11.16, for example, shows the plot of T_∞ obtained by measuring the conductivity of a completely reacted sample of acetic anhydride undergoing temperature ramping. In this case a linear correlation can be used to define the change of T_∞ with temperature.

Having established the rate expression, the experimental conditions to be used, and the calibration of the analytical system, the kinetics of this reaction were used as a test of temperature scanning methods in various reactor configurations.

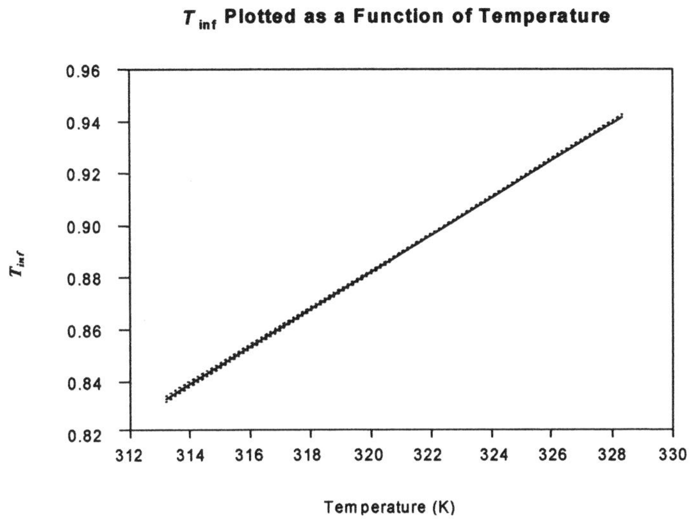

Figure 11.16
An example of conductivity of a reactant mixture at full conversion as a function of temperature.

Kinetics in a TS-CSTR

Temperature scanning methods allow for easy alteration of reactor configuration. All that needs to be changed is the reactor vessel and its associated temperature control, and a different reactor mode becomes available. The CSTR configuration in the setup shown in Figure 11.15 was achieved by replacing the reactor coil shown in the thermostatic bath in Figure 11.15 with a closed stirred vessel. In fact, for purposes of repeated changes between configurations, the CSTR vessel was permanently co-installed with the PFR in the thermostatic bath and the simple switching of a three-way valve redirected the feed from the PFR to the CSTR and back.

The relationship between conversion and conductivity developed above was used to interpret data from the CSTR. As described in Chapter 5, in interpreting data from a CSTR we must take into account the unsteady-state reaction conditions that prevail as temperature is ramped. Even for the case of zero volume expansion, the design equation pertinent to TS-CSTR operation has to account for the constantly changing operating conditions:

$$-kC_A = \frac{C_{A0}}{\tau}X + \frac{dX}{dt}C_{A0} \tag{11.41}$$

Substituting $X=1-C_A/C_{Ao}$ and $\tau = V/v$ (where V is the reactor volume, while v is the volumetric flow rate through the reactor) and dividing both sides by C_A we get:

$$k = -\frac{1}{(1-X)}\left[X\frac{v}{V} + \frac{dX}{dt}\right] \qquad (11.42)$$

where t is the clock time during the process of temperature ramping. We have previously established that

$$X = \left(\frac{T_t - T_0}{T_\infty - T_0}\right)^2 \qquad (11.43)$$

and thus k may be directly calculated from conductivity measurements by substituting this relationship into the expression for k.

Notice that, in the case of the TS-CSTR, the calculation of the derivative dX/dt cannot be done in real-time due to noise in the measurement of T_t and consequently in the calculated values of X. This problem is evident in Figure 11.17 where the original T_t data is shown as a function of time. Figure 11.18 shows the consequent noise in the conversion vs. time plot. Slopes taken point-by-point from such data amplify this uncertainty and are not acceptable for further calculations. To deal with this we need to filter all the data, after the experiment is completed, using a simple 1-dimensional filter. A moving window filter was used in this work.

Since many hundreds of readings were taken during a single ramp, and the expected behaviour was fairly linear, the raw data was filtered using a relatively wide 20-point window. This was re-applied several times until a smooth curve was attained for the conversion versus time curve. The smoothed data is shown in Figure 11.19.

From the filtered conversion values, dX/dt was calculated numerically by using a simple difference equation from point to point. Due to the smoothness of the filtered data, the resultant derivative curve, shown in Figure 11.20, was also quite smooth. Figure 11.21 shows conversion in the CSTR, calculated using the above equations, as a function of outlet temperature, as temperature ramping took place during the TS-CSTR run. The control of temperature ramping was obviously not too precise, nor did it need to be, as we will presently see. The corresponding Arrhenius plot is shown in Figure 11.22.

From the linear regression performed on the data shown in Figure 11.22, the following kinetic parameters were obtained:

pre-exponential factor (A) : 1.87×10^3 l $mol^{-1}s^{-1}$
activation energy (ΔE_a): 45.6 kJ mol^{-1}

The remaining noise and drifts on the Arrhenius plot are the fault of the very simple experimental equipment used. The noise is principally attributable to entrained air bubbles in the feed, which were visible entering the reactor during the experiment. The bubbles disturbed the measurement of conductivity by passing over, or settling on, the conductivity electrodes. Even then, the resultant scatter in the Arrhenius plot is small in comparison with that encountered in the Arrhenius plots from standard isothermal reactors, as is evident from results reported in the literature by various authors.

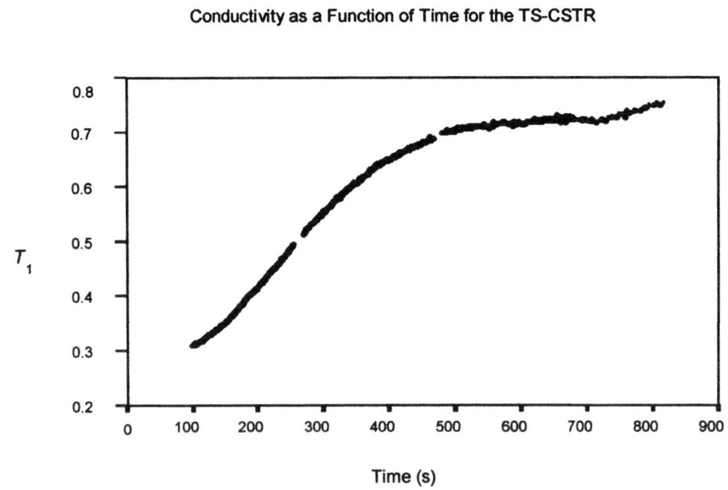

Figure 11.17
Instantaneous conductivity as measured at the outlet of the TS-CSTR during temperature ramping.

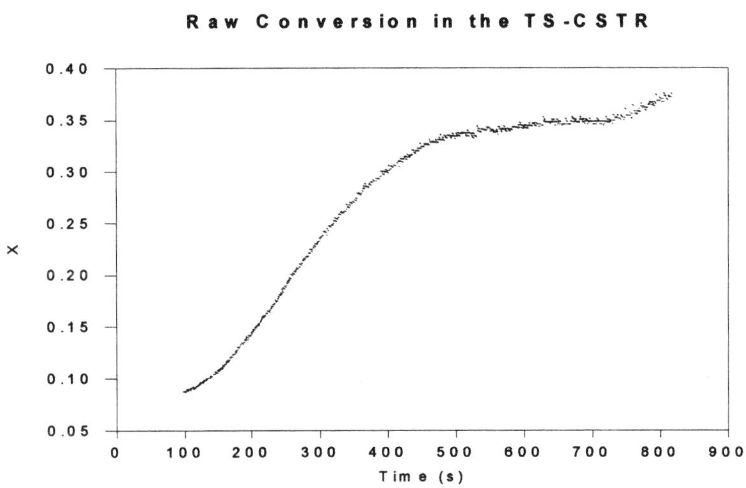

Figure 11.18
Conversion of acetic anhydride, calculated from measurements of instantaneous conductivity as a function of clock time. The space time was kept constant but the temperature was ramped during the run.

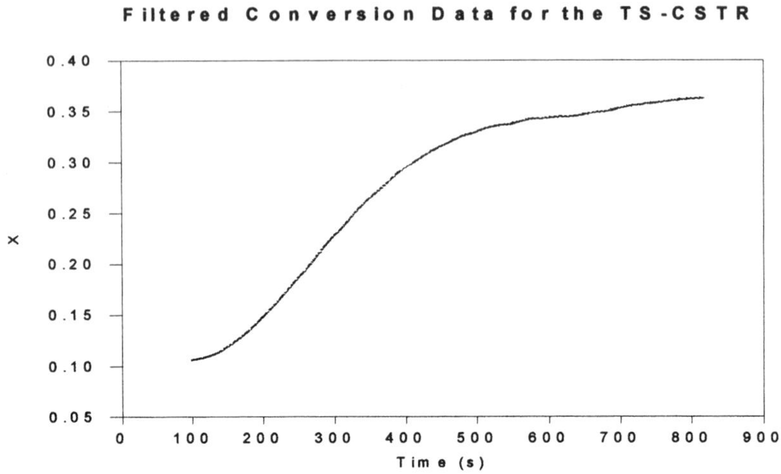

Figure 11.19
Results shown in Figure 11.18 after the application of a one-dimensional filter.

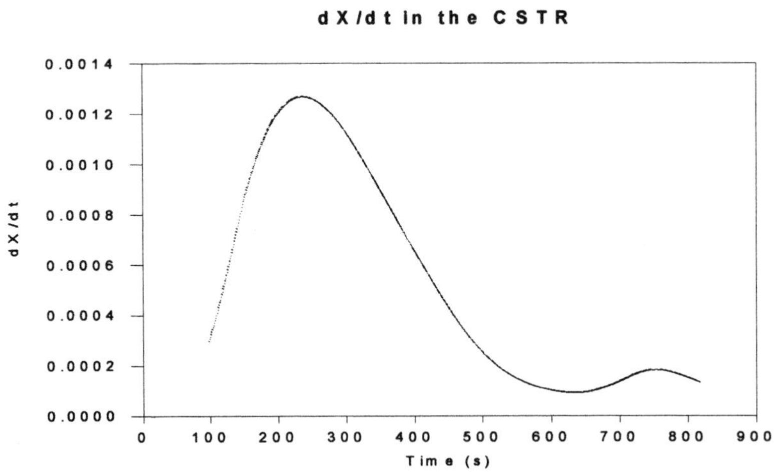

Figure 11.20
Rate of change of conversion with clock time as a function of clock time. This calculated value is necessary to extract rate constants from TS-CSTR data, as described above.

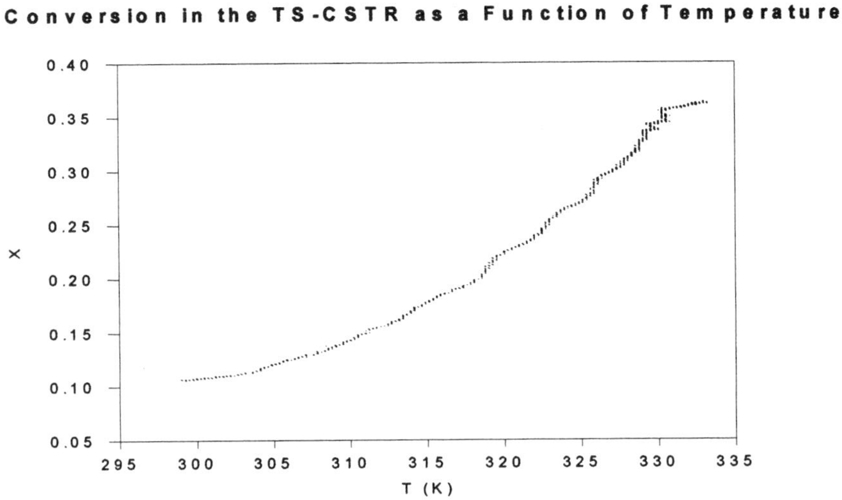

Figure 11.21
TS-CSTR conversion data from Figure 11.18 remapped in the reaction phase plane.

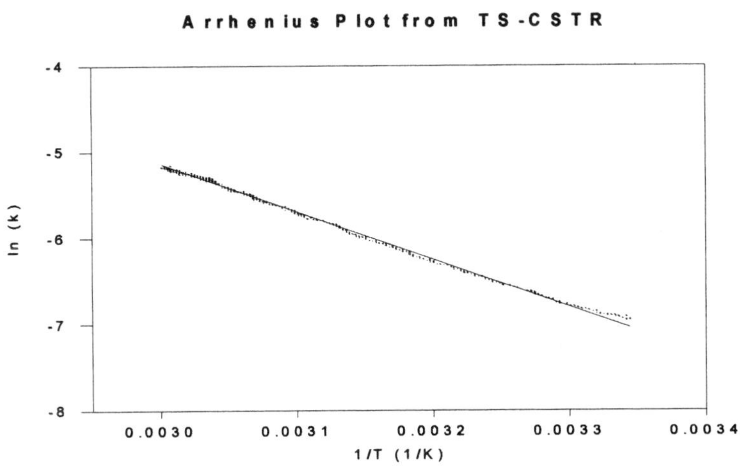

Figure 11.22
Arrhenius plot of rate constants extracted from TS-CSTR data as temperature was ramped during a single run. The number of points available is such that the experimental data produces the almost-continuous line shown. The fitted Arrhenius line is shown as a solid line.

Kinetics in a Plug-Flow Reactor

The configuration used was that shown in Figure 11.15 with about 10 meters of 1/4-inch of plastic tubing serving as the reactor. The shortest tau values were limited by the pressure drop available from the feed pumps. The limitations of treating TS-PFR data using integrated rate expressions were discussed at the ends of Chapter 4 and Chapter 5. In keeping with the requirements listed there, the reaction in this case is essentially thermo-neutral, so that there is no axial temperature gradient.

In that case, as long as τ is small in relation to the time it takes for a one degree rise in temperature, one can calculate the one rate constant of any power-law rate expression by having the data logging program calculate X from the measured T_t and the stored regressions for T_∞ and T_0. In this way a plot of X vs. T, the reactor temperature acquired at the same clock time as T_t, can be generated. Figure 11.23 was constructed in this way in real time, as the experiment proceeded.

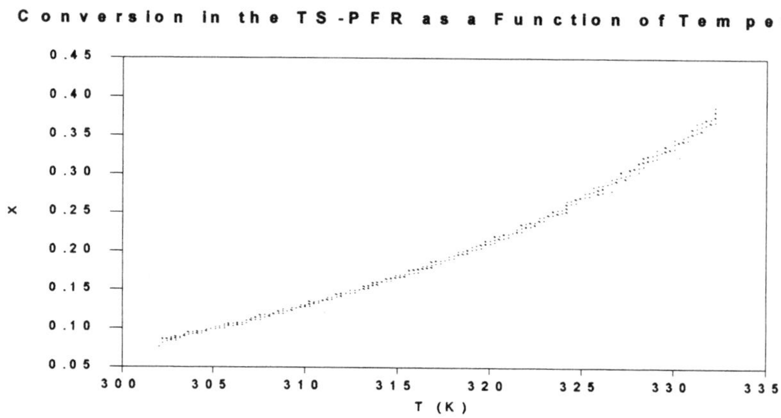

Figure 11.23
A real time plot of conversion vs. exit temperature for a quasi-isothermal TS-PFR

Similarly, Figure 11.24 shows the Arrhenius plot constructed in real time from conductivity measurements by calculating the rate constant k using the relationship:

$$-k = \frac{1}{\tau}\ln\left[\frac{(T_\infty - T_0)^2 - (T_t - T_0)^2}{(T_\infty - T_0)^2}\right] \qquad (11.44)$$

In examining the above figures, one can spot the discrete increments both in conversion (calculated from real-time conductivity readings) and in temperature. Part of the visible scatter is due to the limited 12-bit digital resolution of the data acquisition system used. A more sophisticated data logging board will reduce this form of scatter.

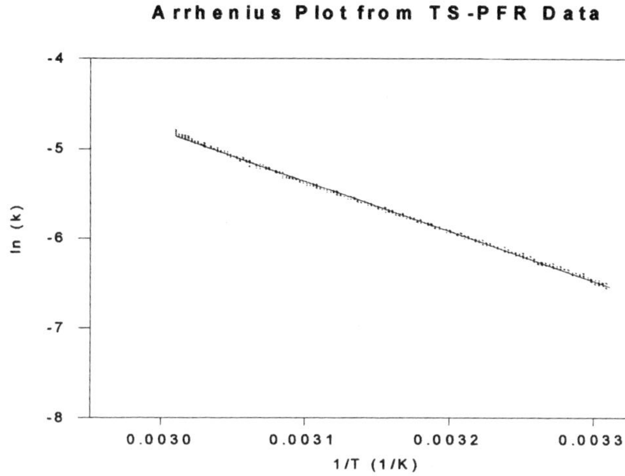

Figure 11.24
Real time plot of the Arrhenius dependence from TS-PFR data. The many experimental values and the best-fit solid line are almost coincident.

The appearance of noise due to poor control capabilities, or to the digital dithering seen in these figures, once more raises an issue of some importance to those equipping a laboratory. All TSR methods produce massive amounts of data. As noted before, this offers an opportunity to employ sophisticated mathematical techniques to clean up the errors. The user has the opportunity to balance the cost of high quality equipment against the effort of sophisticated data processing. In designing a temperature scanning reactor it is always possible to use cheap and fast mathematical data proceeding techniques to compensate to some extent for the use of inexpensive, and therefore less capable, hardware. Notice also that the occasional "wild" point appears on the figures above. These can be edited out of the data-set with a clear conscience in view of the dominance of the well-documented trend of adjoining points. Wild points are easy to spot in this setting and editing will reduce the confidence limits on the parameters obtained without doing violence to the integrity of the results.

Using the above Arrhenius plot from a TS-PFR, a linear regression was performed and yielded the following kinetic parameters:

pre-exponential factor (A) : 2.78×10^3 l g-mol^{-1}s^{-1}
activation energy (Ea): 45.6 kJ mol^{-1}

Kinetics in a Batch Reactor

The BR consisted of the same vessel used as the CSTR but with no flow. The vessel was charged with an appropriate volume of water, the electrode immersed in the water, and stirring initiated. At time t = 0 the desired volume of acetic anhydride was injected into the vessel. Figure 11.25 shows conversion of the acetic anhydride in a batch reactor as it proceeded as a function of clock time. Using clock time (t) instead of space time (τ), the reaction rate constants were calculated in real time using the equations and methods given above for the TS-PFR. The kinetic parameters were determined using a linear regression on the Arrhenius plot of rate constants calculated in this way.

pre-exponential factor (A) :	1.82×10^3 l g-mol^{-1}s^{-1}
activation energy (Ea):	45.2 kJ/mol

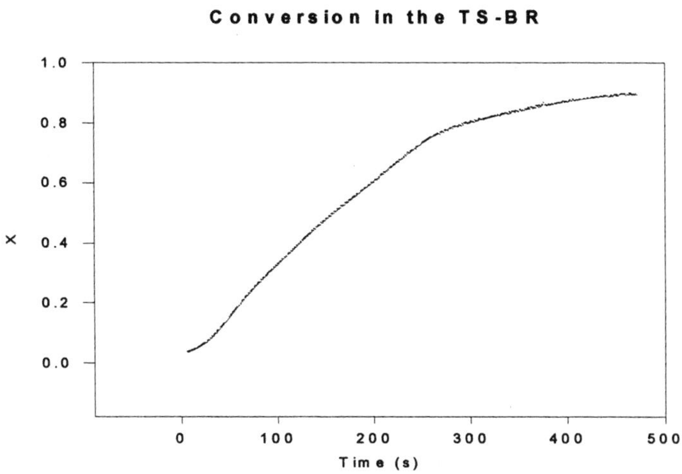

Figure 11.25
Conversion vs. clock time in the TS-BR. The BR configuration was achieved by filling the reactor with water and at time zero rapidly injecting the required quantity of acetic anhydride.

Table 11.3 shows a compilation of kinetic parameters determined using the three types of temperature scanning reactor and compares them with corresponding parameters found in the literature. We see that there is much agreement as to the activation energy of the reaction. Any disagreements regarding the rate parameters concern the value of the pre-exponential, and even there, most authors tend toward the lower value found using the TSR. Perhaps the reports with high values for the pre-exponential contain an unrecognized systematic offset caused by an extraneous constant factor.

Table 11.3.
Experimental Activation Energies and lnA for the Hydrolysis of Acetic Anhydride

	Activation Energy (kJ/g-mol)	ln A (l/g-mol/s)
Asprey et al.		
TS-PFR	45.6	7.93
TS-CSTR	45.6	7.54
TS-BR	45.2	7.51
TSR studies average	**45.4**	**7.66**
Shatynski (1993)	46.9	12.74
Bisio and Kabel (1985)	46.5	12.80
Glasser and Williams(1971)	45.2	7.95
Cleland (1956)	44.4	7.80
Eldridge (1950)	43.8	7.53
King and Glasser (1965)	39.8	9.93
Published average	**44.4**	**9.79**

Considering the varied and often difficult methods used for tracking conversion in this system, these differences in kinetic parameters come as no surprise. On the other hand, considering the simplicity of the kinetics in this reaction, the discrepancies between the reported parameter values give warning of the uncertainties that may lurk in reports of kinetic parameters for more complex systems.

Improved experimental methods in kinetics will eliminate most instances of unexplained discrepancies in published experimental results.

Variants on the Methods of Data Interpretation

The method of integral data interpretation described above is the most straightforward in this case and offers real time data interpretation. It is fast, engaging and useful as a rapid means of determining the Arrhenius parameters. In line with our timeless desires, it offers instant gratification. However, it is less general than the standard processes of filtering and sieving of TSR data described in more detail in Chapters 7 and 11. It is worthwhile at this point to consider what other methods of data interpretation for rate studies are available.

A more general method of using integrated forms involves using the standard procedure where one waits until the TSR experiment is completed and then proceeds to smooth the (X, T, τ) surface as described in previous discussion. On the smoothed surface one identifies the operating lines for the system and uses them in an integral method of data interpretation. This removes the restriction employed in the hydrolysis study, that the system be at quasi-isothermal conditions, and makes the method more general.

From the smoothed surface one can select any operating line, just as was done to obtain rate data in the case of carbon monoxide oxidation and the steam reforming of methanol. In the case of integrable rate expressions there is no need to take the next step of calculating rates along the operating lines. Instead, the operating line supplies two plots: an X vs. τ plot and a T_e vs. τ plot. The conversion (X) data along the operating line is fitted with an integrated rate expression whose constants are expressed as functions of temperature. The conversion itself is a function of tau and the temperature dependent constants. This method therefore requires us to iterate (do a numerical integration) with tau and temperature. What we need to find is the best set of Arrhenius parameters to calculate the observed behaviour of X for an increment of reactant as it proceeds along this operating line. The final optimization procedure involves arriving at one optimum set of Arrhenius parameters after fitting a selection of such paths along various operating lines.

On reflection, it is obvious that the avoidance of slope-taking has its price in computational difficulties and in the restriction of this method to rate expressions that have an integrable form. Considering the painless way that slopes (rates) are calculated using the methods described in Chapter 8, this seems too high a price to pay. The interpretation of TSR data using integral methods has little to recommend it in most cases. Integral methods have been attractive in treating the sparse data from conventional isothermal operation for specific reasons that do not apply to TSR data. Their attraction has been based on the opportunity to make a linear Arrhenius plot of the sparse set of isothermal parameter estimates available from a few conventional isothermal measurements. Since a straight line is easy to draw by eye, and linear regression is familiar to most, data interpretation is facilitated by this approach. On the other hand, this method restricts much kinetic interpretation to rate expressions that have a closed form integral.

The main attraction of using the integral approach in conventional studies is that it avoids the need to measure rates of reaction. Instead, the output conversions from several isothermal runs at different space times are plugged into the integrated rate expression and the rate parameters are optimized as simple constants for a given temperature. Since there usually are few parameters in a rate expression, a few runs will suffice to define all the constants at one temperature; certainly, fewer runs than would be necessary to make valid estimates of rates from a plot of X vs. τ. Repeating this procedure at several temperatures yields a set of constants suitable for plotting on an Arrhenius plot. This procedure minimizes the number of isothermal runs necessary to obtain the rate parameters.

However, the advantages of this method pale by comparison when one considers the wealth of data readily available from a TSR. Moreover, the integral procedure misses out on the powerful armory of mathematical methods of data filtering and correction outlined earlier in this text. Finally, TSR procedures can inherently deal with all forms of rate expression, integrable or not. This by itself presents a surprising liberalization in the choice of rate expressions that can be considered and of systems that can conveniently be studied.

It must be emphasized that one cannot simply sieve out isothermal sets of (X, T, τ) from TSR experimental data and treat them with isothermal forms of the integrated rate expression. Such data does not lie along an operating line and therefore does not represent the fate of any plug of reactant passing through this, or any other possible, reactor.

Experimental Issues in TSR Operation

In the above study of acetic anhydride hydrolysis, and in some other systems, TSR techniques can be implemented to yield kinetic parameters in real time, greatly increasing the gratification factor as interpreted data is made available even as the experiment proceeds. Clearly, this is possible only if rate data can be obtained in real time during a run, for example if the integral form of the rate expression can be interpreted to yield a rate constant in real time. Even then, one must have a previously established kinetic rate expression available for data fitting. In most cases, however, the rate data does not become available until after completion of the TSR experiment. These differences notwithstanding, both in the instant gratification cases and in all other TSR studies, experimentation entails first establishing various calibrations and collecting various ancillary information before interpretable data can be gathered.

In the hydrolysis reaction for example, this ancillary information entails preliminary runs that establish T_o and T_∞ values over the temperature range to be traversed by the rampings. In adsorption studies, as a very different example, ancillary studies entail establishing the value of the monolayer. Yet again, in catalytic studies there is a need to calibrate sophisticated analytical procedures and to establish that the data will be collected in a diffusion-free regime. And in all cases there is the need to periodically calibrate flow meters, thermocouples, pressure sensors and so on.

The set-up activities associated with a TSR experiment normally take more time than that required for the completion of a subsequent experiment. It is therefore advisable to formulate a plan for a complete set of investigations on a given system before the equipment is assigned to a new study. The plan must include the immediate processing of acquired data and its fitting to candidate rate expressions.

Once the candidate rate expressions are reduced to some small number, a search for interesting reaction conditions should be undertaken by simulation with each of the remaining candidate rate equations,. On the basis of the results of this search, further experiments designed to distinguish between rival rate formulations can be immediately undertaken. Finally, on the basis of all the experiments, an all-up fitting of all data collected is made to refine the rate parameters of the chosen rate expression. The final rate expression is then available as a broadly applicable descriptor of the kinetics of this system, and the equipment can now be recalibrated for other work.

In catalyst development and in academic catalytic studies, subsequent samples of similar catalysts will normally operate via the same mechanism of reaction. In that case the acquisition of rate parameters for subsequent samples of catalyst is reduced to simply confirming that the experiment was carried out in a region where a good estimate of the parameters can be obtained and fitting data from the new catalyst samples to the previously established rate equation. After a sufficient number of samples have been studied in this way it should be possible to correlate changes in parameter values with changes in catalyst properties.

> *We come here to the ultimate goal of kinetic studies, and the benefits of temperature scanning experimental methods in pursuit of that goal. The goal of kinetic studies is to investigate rates of reaction over a broad range of reaction conditions so that a unique and all-encompassing rate expression, preferably one corresponding to a plausible reaction mechanism, can be found. Using temperature scanning methods it is possible to obtain enough data to relate catalyst properties to fundamentally meaningful kinetic parameters of the rate expression of each of the samples tested.*

The understanding of reaction mechanisms and what such a development will afford cannot be overestimated. This information will allow the *rational* design of catalysts, as the kinetic effects of changes in formulation are understood in terms of adsorption phenomena, other rate processes, and the related changes in catalyst properties. Simulation, using the established rate expression, can then be used to find preferred sets of parameters. On the basis of the correlations between catalyst properties and rate parameters this would indicate the desired catalyst formulation. It would then be up to the chemist in charge of synthesizing new catalyst formulations to prepare the desired composition.

Kinetics can be a powerful a tool for understanding reaction mechanisms. In turn, understanding the mechanism and its rate expression allows us to engineer the reaction by influencing elementary steps in the overall conversion process. With the new and highly productive experimental approaches and techniques described here, kinetic studies no longer need to be a tedious undertaking, requiring many runs before sufficient data became available to verify a mechanistic postulate. This accumulation of kinetic understanding is the kind of quantitative work that must be done if we are to unravel the kinetics and mechanisms of reactions. A successful accumulation of the pertinent rate parameters and their correlation with material properties will surely be of great practical use in the development of new catalysts and in the improved of reactors operating at industrial reaction conditions.

12. Using a Mechanistic Rate Expression to Understand a Chemical Reaction

Advances in science proceed in well defined steps, from observation through correlation to interpretation. Finally, at the end of the quest, lies a reliable understanding of the system beyond the few points directly observed. This level of generalization makes it possible for us to manipulate the physical world to our advantage.

A Plan of Action

The principal issues connected with experimental studies in chemical kinetics have been covered in Chapters 1 to 11. Once the mechanistic rate parameters are established by a kinetic study, the reaction is ready to be examined under various conditions using simulation. This is the stage at which we can see the advantage of mechanistic interpretations in understanding the physical events underlying experimental observations and in making predictions based on such understanding. Mechanistic equations afford a reliable way of generating mappings of reaction behaviour that will lead to this understanding, as well as being useful in more utilitarian applications, such as the optimization of commercial reactor design and control.

In undertaking the search for this level of understanding, the first issue one faces is the identification of graphical views of the reaction that will be most helpful in making sense of its behaviour. Most traditional graphic representations of kinetic behaviour are two dimensional and here we will follow that convention., In time, however, three dimensional views of the kind shown in previous chapters may come to dominate data interpretation. Even in two dimensions there is a large array of mappings of reaction space made possible by combinations of the many reaction variables taken two by two. In practice not all such mappings need to be examined.

To help draw an adequate picture of the system, some auxiliary information may also have to be obtained by experiment or retrieved from the literature. Much useful and pertinent information is available in standard form in handbooks, such as heats of reaction, heat capacities, heat transfer coefficients and so on. There is little debate as to the correctness of the values found in this way, although the units used in various sources can differ and must be carefully noted and converted to some common system of measures before interpretation is attempted. Other required information might involve auxiliary experimental work on adsorption, diffusion, physical stability, activity decay and other sidelights on the system.

Although these procedures are generally applicable to all chemical reactions, in the following pages we use the simulated behaviour of the catalytic oxidation of carbon monoxide as an example to map and understand the peculiarities of this system. We also search for the best reaction conditions for eliminating carbon monoxide from various streams under certain constraints. The rate expression we will be examining is the same as that previously presented in Chapter 11.

$$-r_{CO} = -\frac{dP_{CO}}{d\tau} = \frac{k_r K_{CO} K_{O2}^{1/2} P_{CO} P_{O2}^{1/2}}{(1 + K_{CO} P_{CO} + K_{O2}^{1/2} P_{O2}^{1/2})^2} \qquad (12.1)$$

In practice, experimental data that follows this form of behaviour might at first have been fitted using a generalized empirical equation containing three temperature-dependent constants and two temperature-independent constants:

$$-r_{CO} = \frac{k_1 P_{CO} P_{O2}^n}{(1 + k_2 P_{CO} + k_3 P_{O2}^n)^m} \qquad (12.2)$$

A good fit of data to this equation should stimulate a search for the mechanism(s) that can generate this form of rate expression. Once a mechanistic rate equation of this form, say that in equation 12.1, is identified as the appropriate one for a given system, the three fitted constants k_1, k_2 and k_3 can be untangled to yield the three fundamentally meaningful constants k_r, K_{CO} and K_{O2}. At this point we should recall that equation 12.1 is a mechanistic rate expression that comes from a more fundamental form:

$$-r_{CO} = k_r \theta_{CO} \theta_{O2} \qquad (12.3)$$

This will lead us to take a closer look at the more fundamental question of the behaviour of the fractions of surface covered by the two reactants, θ_{CO} and θ_{O2}. Comparing equation 12.1 with 12.3 we see that:

$$\theta_{CO} = \frac{K_{CO} P_{CO}}{1 + K_{CO} P_{CO} + K_{O2}^{1/2} P_{O2}^{1/2}} \qquad (12.4)$$

$$\theta_{O2} = \frac{K_{O2}^{1/2} P_{O2}^{1/2}}{1 + K_{CO} P_{CO} + K_{O2}^{1/2} P_{O2}^{1/2}} \qquad (12.5)$$

Since both fractional coverages are dimensionless, the dimensions of the reaction rate reside in the catalytic rate constant k_r, which is itself a composite parameter containing a term for the total site concentration, S_0, squared:

$$k_r = k_r' S_0^2 \qquad (12.6)$$

where k_r' is the fundamental rate constant for the surface reaction of adsorbed CO with adsorbed O.

Its dimensions depend on the definition of the space time used in the given work, making it very important to report not only the units of tau but also its precise definition (see Chapter 2). Having established which mechanistic rate equations are pertinent to the system we proceed to refine their parameters using experimental data and then map the behaviour of the system by simulation of its many (perhaps un-investigated) features using the fitted rate parameters

Mapping Reaction Rates

Mapping the Effects of Feed Ratio

In the case of the catalytic oxidation of carbon monoxide, an interesting mapping is that of rate of reaction versus conversion at isothermal conditions. The reason is that, as we saw in Chapter 11, certain reaction conditions lead to a maximum in reaction rate at partial conversion. Using the best parameters for the rate expression reported in Chapter 11 we can simulate a variety of reaction conditions and find that at 1 bar total pressure, for a stoichiometric molar feed ratio of $CO/O_2 = 2$, the rate of reaction at various reaction temperatures is predicted to behave in the way shown in Figure 12.1.

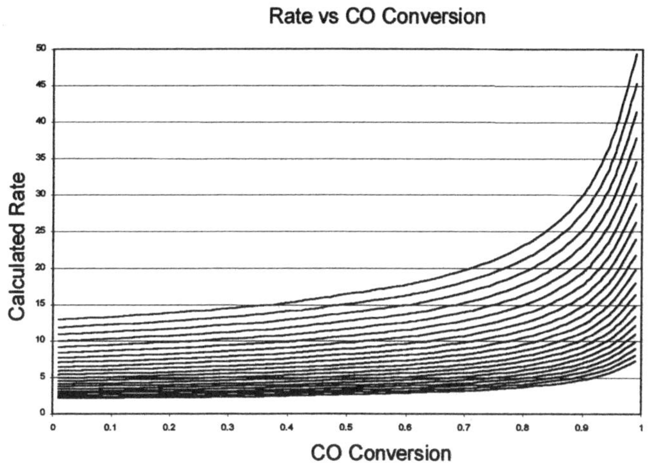

Figure 12.1
Simulated rates of reaction at a feed ratio of 2.0. The upper curve corresponds to 570 K, the bottom to 530 K.

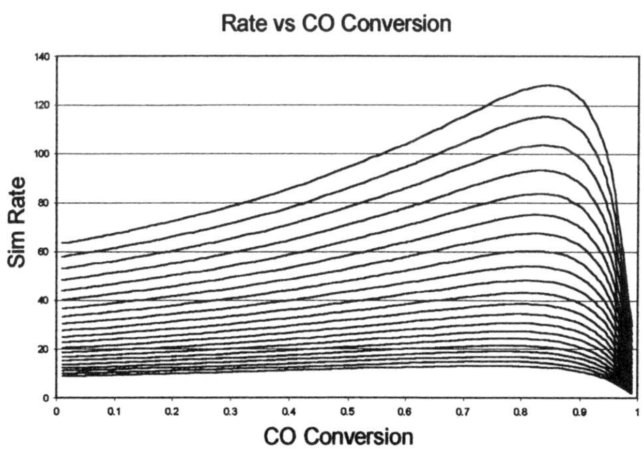

Figure 12.2
Same reaction conditions as in 12.1 but at a feed ratio of 0.2 rather than 2.0.

The surprising observation that a sharp maximum in rate is achieved at some high conversion calls for further thought. One would expect the rate of reaction to decrease as conversion proceeds, since the concentration of the reactants is obviously decreasing. Instead, a detailed examination of the behaviour in Figure 12.1 shows that there is a maximum that occurs at about 99% conversion at all isothermal conditions between 530K and 570K. Moreover, the maximum in rate does not stay at this high conversion under all conditions. Further simulations show that the rate-maximum changes position with feed composition, as in Figure 12.2 where the feed ratio is reduced to 0.2.

There is now a maximum just below 90% conversion but the rates are much higher, while the maxima are much broader. What physical factors underlie this unexpected behaviour? After all, a lower feed ratio implies that the concentration of one of the reactants is reduced while the other is raised, compared to the first example in which the feed ratio was 2. One might reasonably have expected that a stoichiometric mix of the reactants, as in a feed ratio of 2, should have resulted in a higher reaction rate. Clearly this is not so.

In catalysis the reasons for all kinetic behaviour lie in the behaviour of surface coverage by the reactants. This means that the kinetics of certain catalytic reactions - and the catalytic oxidation of CO via a bimolecular surface reaction is one of them - do not depend directly on gas phase concentrations. To understand Figures 12.1 and 12.2 we need to examine the behaviour of θ_{CO} and θ_{O2} as expressed by equations 12.4 and 12.5. It is only by understanding the behaviour of the fractions of surface covered by adsorbed species that an understanding of *any* catalytic reaction can be gained. Since at present there is no way to measure the adsorption isotherms of reacting species at high temperatures, we need a reliable mechanistic rate expression to examine this aspect of the reaction. An appropriate mechanistic rate expression will permit us to reliably simulate the behaviour of the isotherms *from kinetic data.*

We examine the behaviour of adsorption under reaction conditions using equations 12.4 and 12.5 to simulate various coverages in the catalytic oxidation of CO, as shown in Figures 12.3 and 12.4. The course of the reaction on these two figures is as follows. The rightmost curve in Figure 12.4 shows total surface coverage at a feed ratio of 0.2, starting with zero conversion at the pendant point at about 98% coverage. As reaction proceeds, total coverage decreases, while the rate at first increases and then decreases until zero rate is obtained at 100% conversion, as shown in Figure 12.2.

The total surface coverage, however, does not go to zero at zero rate but ends up at about 89%. At the same time, coverage by CO starts high, about 85% of available sites, and proceeds to decrease to zero on the left as conversion proceeds to 100%.

Finally, coverage by oxygen atoms begins low on the left, about 13% at zero conversion, and increases as reaction proceeds. At 100% conversion the total coverage of 89% of the surface is that due to oxygen atoms alone. All the CO has been converted and none is left to cover the surface. The behaviour on Figure 12.3 is similar to that in Figure 12.4. Figure 12.3 shows the coverage of the surface for the feed ratio of 2.0 as reaction proceeds at 570 K. The highest rate is achieved at about 80% surface coverage when coverage by CO and O is respectively about 70% and 10%. This is a situation where many adsorbed CO molecules find themselves without an oxygen atom on a proximate site. This is clearly undesirable, since our understanding of catalytic surface reaction mechanisms tells us that a bimolecular reaction cannot take place in the absence of such proximity.

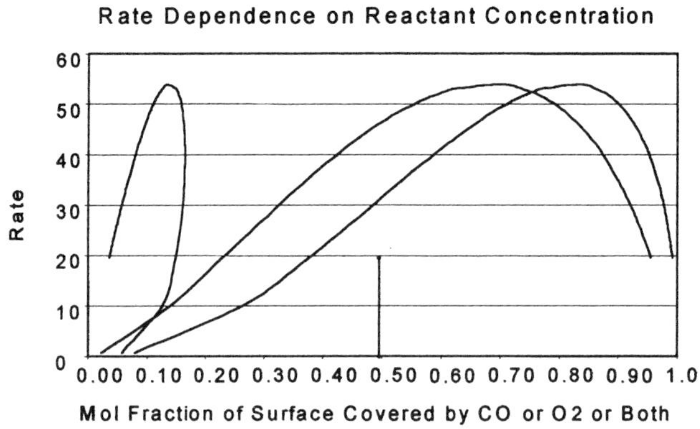

Figure 12.3

The curve starting furthest to the right shows total coverage of the surface by adsorbed species as reaction progresses. Next to it and to the left is the curve for coverage by CO. This curve also starts on the right. The isolated curve on the left of the field shows coverage by O. This curve starts on the left. The coverages are seen as they evolve with conversion from these starting points. In the end, when conversion goes to 100%, all coverages for a stoichiometric feed ratio of 2.0 go to zero.

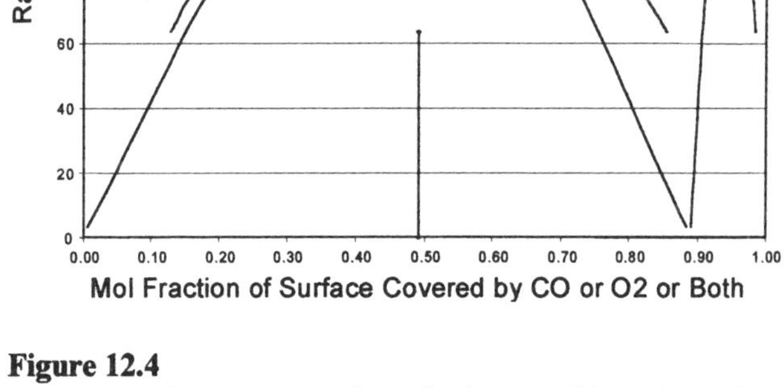

Figure 12.4

Rate vs. surface coverage for a feed ratio of 0.2. The isolated curve on the right shows total coverage. The curve ending near the origin shows coverage by CO. The remaining curve shows coverage by O. The direction of change with conversion is the same as in Figure 12.3.

Compare this situation with that on Figure 12.4 at a feed ratio of 0.2. There we see that total surface coverage is somewhat higher, 97% at the rate maximum, but the notable feature is that coverage by CO and O is almost equal at maximum rate, about 52% and 45% respectively. This means that almost all available active sites are covered (total coverage at that point is 97%) and therefore, on average, there will be a CO and an O adsorbed on adjacent sites. The short vertical line in Figures 12.3 and 12.4 shows the point at which the total coverage would be split equally between the two adsorbed species.

In principle the absolute maximum rate of a bimolecular reaction proceeding by this (Hinshelwood type) mechanism is achieved at 50% coverage by each of the reacting species and 100% total coverage. Our intuition that maximum rate occurs for a 50/50 mixture of reactants is correct, but the reactants in this case are surface species, not their gas-phase precursors. The condition of 1 bar and a 0.2 ratio of CO/O_2 therefore yields almost the maximum possible rate for this reaction, on this catalyst, at this temperature. Further optimization of the feed ratio and reaction pressure is possible but there is usually little to be gained since the maximum rate in this region of reaction conditions changes very slowly.

Figure 12.5 shows a different mapping of the behaviour of surface coverage with conversion, at 1 bar, for a feed ratio of 0.2. We see that, at low conversion, coverage by O is low while that by CO is high. As conversion increases, coverage by O increases while that by CO decreases. At about 88% conversion the two curves cross. The maximum rate is found near the crossing point of the two coverage curves where the curve for the product of the fractions covered, ($\theta_{CO} \times \theta_O$), shows a maximum. On Figure 12.5 this takes place at about 85% conversion.

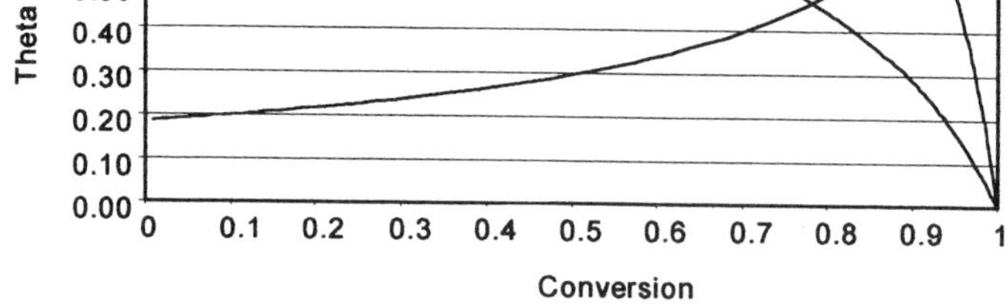

Figure 12.5

Total coverage is shown by the uppermost curve. The next curve on the left shows CO coverage decreasing from 80% to zero at full conversion. The lowest curve shows O coverage increasing from 18% to the point where it is the sole species on the surface, at complete conversion. The curve with the maximum shows the product of the two coverages multiplied by 4. This shows to scale the approach of the product of the two fractions of surface covered to the maximum possible product of ¼, and so locates where the maximum rate will be found.

Throughout the range of conversions in Figure 12.5, total coverage decreases with increasing conversion, but not as much as one might expect. Losses of CO coverage are made up by the competing species O since at this feed ratio the concentration of its gas phase source, O_2, increases with conversion relative to CO.

The location of the maximum in the rate continues to drift toward zero conversion until, at a feed ratio of about 0.03, the maximum rate is found at zero conversion, still at about the same absolute value of rate as it had at a ratio of 0.2. At lower feed ratios at this temperature the rates are always highest at zero conversion and decrease monotonically toward zero at 100% conversion. Throughout much of this range, total coverage remains above 90%.

The range of the parameter describing feed ratios is therefore divided into two distinct regions. In the region below a feed ratio of 0.03 the maximum rate occurs at zero conversion and decreases with decreasing feed ratio. Above this ratio there is a maximum in the rate of reaction at partial conversion. In this region the maximum rate moves to lower rates and higher conversions as the feed ratio increases. Two investigators, each working in one of the two regimes, will report very different behaviour for this reaction. Their limited observations will naturally lead to very different interpretations of the kinetics and mechanism of the reaction. The resultant confusion can only be eliminated by a broad-ranging study of the reaction under a variety of reaction conditions.

At the same time as the above changes in rate behaviour take place, the proportion of surface covered by each of the two species, CO and O, invariably changes monotonically with decreasing feed ratio from that where the majority of the surface is covered by CO, at high ratios, to that where the majority is covered by O, at low ratios. The changeover from dominance by CO coverage to dominance by O coverage at 570 K takes place at a feed ratio of about 0.03. All this takes place without major changes in total coverage. Total surface coverage at maximum rate drifts up to about 95% at a ratio of 0.03 from about 90% at a ratio of 2.

We can sum up these observations by noting that, at a constant temperature, coverage by the two surface-resident reactants CO and O changes with feed composition. This comes about due to the different orders of dependence of the adsorption equilibria for the two species. While CO adsorbs molecularly, and its coverage of the surface is dependent on the first power of its gas phase concentration, that of O involves dissociative adsorption of gas phase O_2. This makes surface coverage by O dependent on the square root of the gas phase concentration of O_2. The result is that the proportion of the surface covered by the two reactants varies with feed composition, even under isothermal conditions. At each temperature there will be a specific feed ratio where the two surface species are present in almost equal amounts. At that condition the rate of reaction will be maximized. Nearby feed ratios will show similarly high reaction rates.

Due to the fact that gas phase reactant consumption is in a ratio of $CO/O_2=2$, all feeds other than that with a ratio of 2 will result in reactant composition ratio changing with conversion. This we know from stoichiometry. Our mechanistic rate expression tells us more. The fact that adsorption of the two reactants shows different dependencies on their gas phase concentrations should alert us to the possibility of unusual kinetic behaviour within the range of reaction conditions being contemplated. Even at a feed ratio of 2, it turns out, there is a maximum in rate at about 99% conversion due to differences in the isotherms for the two reactants and the fact that in the end all the feed is used up causing the surface coverage and the rate to go to zero. Coverage by O actually increases slowly up to about 99.8% conversion. Similar phenomena appear in all catalytic reactions with similar mechanisms and stoichiometry. Unusual effects can also arise if adsorbed but non-reacting components are present in the system and appear in the denominators of the isotherms.

As a by the way, and to make a point, we can simulate the kind of data one might gather using conventional isothermal data collection methods. A plot of conversion versus space time at 530K, 1 bar and a feed ratio of 0.4 is shown in Figure 12.6. Such data is normally acquired in conventional isothermal kinetic studies using a PFR. The number of points shown in Figure 12.6 is probably larger than that available in the average investigation and this scarcity of data will contribute to the uncertainties that can result from an examination of such sparse and unfiltered data.

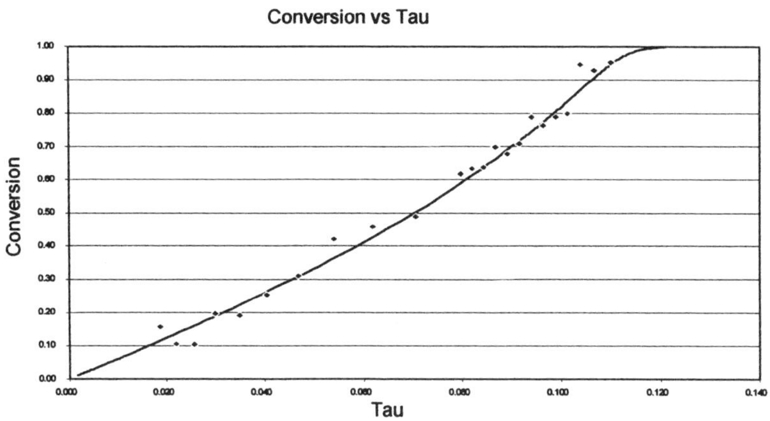

Figure 12.6
Conversion as a function of space time at 530 K and a feed ratio of 0.4. Such a curve would be delineated in an isothermal study by individual experiments involving runs at a series of space velocities. The points show experimental results with scatter similar to that discussed in Chapter 2.

Experimental points, such as those shown, may well give one pause. Data with minimal noise, such as that shown by the solid line, would correctly reveal an increase in rate with increasing conversion and a sharp slowdown in rate at high conversions, so that the last increments of conversion require an anomalously large increase in space time. This information would surely encourage serious thought. However, if the data were noisy (3% random error has been included in the points shown in Figure 12.6) one could easily be misled and might even see the increase in conversion as being linear with space time. Such results, when interpreted in the commonly-offered word images, might lead to explanations such as: "zero order reaction", "acceleration by products", "auto-catalysis", or "poisoning by high levels of products", the latter put forward to account for the sharp drop in rate at high conversion. None of this is true, and the real situation emerges only when one fits an appropriate mechanistic rate expression to an adequate collection of noise-free data.

None of this insight is available from empirical rate equations. It should also be pointed out that all the conclusions reported here are based on a rate expression fitted to non-isothermal data obtained using a TSR. This is the essence of satisfactory data correlations: one can examine reaction conditions that were not accessible by direct experimentation. Such procedures are an uncertain business in the case of empirical rate expressions. On the other hand, behaviour simulated using appropriate mechanistic rate expressions can be safely examined under any reaction condition.

Mapping the Effects of Total Pressure

Increases in total pressure generally result in moving the maximum in rate to higher conversions. For example, rather than the maximum sinking into the y-axis at a feed ratio of about 0.03, as it did at 1 bar, at 5 bar this event takes place at a ratio of about 0.01, while at a feed ratio of 0.03 there now appears a maximum in rate at about 60% conversion.

At the same time, the absolute maximum rate is the same as it was before, regardless of total pressure. This is clear from the fundamental expression given in equation 12.3, where we see that k_r is a function of temperature and site concentration but independent of reactor total pressure. At the same time, the maximum product of the two θ fractions describing surface coverage, whose sum cannot exceed 1, is ¼. Thus for each temperature there is an absolute maximum rate available at a specific feed ratio and pressure.

There remains the question: why does the maximum rate at a given feed ratio move to higher conversion at higher total pressure? Again the answer lies in the behaviour of the surface coverage. Examining equations 12.4 and 12.5 we note that, as total pressure increases while the feed ratio remains the same, the partial pressures of the two reactants will change in the same proportion. The result is that, regardless of the denominators in equations 12.4 and 12.5, θ_{CO} will increase relative to θ_O. This in turn causes surface coverage to move away from a condition such as that shown in Figure 12.4 toward that in Figure 12.3, with a resultant decrease in rate.

Another way of saying the same thing is that higher pressures at a given feed ratio result in an increased proportion of the surface being covered by CO. In cases where this trend results in departure from, rather than approach to, a condition of 50/50 coverage by the two reactants, a pressure increase decreases the maximum rate at a given feed ratio.

> *Total pressure beyond a certain value does not alter the maximum achievable rate of reaction in this type of mechanism. Below that pressure, and if total surface coverage is already high at the pressures under consideration, increases in pressure can result in a decreased reaction rate at a given feed composition.*

Mapping the Effects of Temperature

The effects of temperature on this and similar reactions are very complex and highly dependent on the relative sizes of the energy terms in the rate parameters. Only one thing is clear about the effect of temperature: the catalyst is used most efficiently when operating isothermally at the highest temperature accessible. In the example we are considering, the temperature sensitive parameters are:

activation energy of the reaction:	ΔE_r	= 1.46E+05
activation energy for CO adsorption:	ΔH_{CO}	= −7.53E+03
activation energy for O adsorption:	ΔH_{O2}	= −8.30E+04

We see from equations 12.1 and 12.3 that the net activation energy for most conditions away from the absolute maximum rate lies between:

$\Delta E_{net} = \Delta E_r - \Delta H_{CO} - (½)\Delta H_{O2} = 5.55E{+}04$ and $\Delta E_r = 1.46E{+}05$.

The consequence is that overall rate of reaction always increases with temperature, while total coverage by both reactants decreases. One might also think that coverage by O decreases more rapidly than that by CO, but things are not that simple. While coverage by O does indeed decrease, that by CO increases with temperature, as we will soon see.

The upshot is that maximum rate at a given feed ratio is shifted to the higher conversions as temperature is increased. This also means that the feed ratio at which maximum rate first appears at zero conversion is shifted to lower feed ratios. The absolute maximum rate increases with the highest activation energy possible, ΔE_r = 1.46E+05, making temperature effects greater if one follows operations near the maximum achievable rate by adjusting the feed ratio. This variability in the apparent activation energy can lead to confusion if data is collected using only standard test conditions.

Mapping the Effects of Dilution

Dilution by an inert non-adsorbable additive is equivalent to reducing the reactant pressure without changing the feed ratio. Although there may be reasons why a dilute reactant stream has to be used, the purely kinetic effects of diluents are simple. The novel issues that are raised by the addition of a diluent concern changes in space velocity, the heat capacity of the feed stream, and possibly heat transfer between the reactants and the surroundings. Some of these effects may be advantageous but, since dilution increases the space velocity based on total feed rate, it results in a lower throughput of the reactants at a fixed space velocity.

> *In order to obtain the maximum available reaction rate, one has to operate at zero dilution and adjust the feed ratio, total reaction pressure, and temperature to a specific value within the range of permissible operating conditions.*

Maximizing the Conversion of Carbon Monoxide

All reactor operations are constrained by physical realities. These result in maximum and/or minimum constraints on the operating conditions of a commercial reactor. In the interest of this illustration we will chose the following short list of constraints.

- The reaction temperature cannot exceed 570 K.
- The reactor pressure cannot exceed 5 bar.
- The space-time limits in the PFR are 0.2 to 0.02 sec in volumetric terms.

Within these constraints we can now try to achieve any of several objectives. We will consider only the simplest of them.

Maximize CO Conversion Productivity in a PFR.

Conversion of CO per unit of catalyst volume (we will use tau as a proxy for this) we will call *productivity*. Maximizing productivity can involve several approaches, of which we will consider only two of the simplest of the many possible objectives.

1. *Maximize the conversion of CO in terms of moles converted per unit of catalyst volume while maintaining 99% conversion per pass.*

Mapping of the behaviour of the system under this premise shows that, under all conditions permitted by the above restrictions, the operation should be carried out at the highest temperature allowed. The oxidation of CO is a highly exothermic reaction and allowing the temperature to increase to this maximum from some lower inlet temperature will not increase productivity over that available under isothermal operation at the maximum allowed temperature. In practice this type of operation causes problems since maintaining isothermality in highly exothermic reactions is difficult, but we will ignore this issue. The free parameters one can manipulate are then the feed ratio and the reactor total pressure.

In order to obtain a measure of reactor productivity we calculate the moles of CO converted per mole of total input at the desired conversion using:

moles converted = conversion*[feed ratio/(1+feed ratio)]*(1–dilution ratio)

and divide this number by the space time, tau, of the operation at each condition. This allows us to construct a plot of productivity versus feed ratio. Figure 12.7 shows a mapping of the pertinent space.

Perhaps surprisingly we see that productivity is higher at lower total pressures. This comes about because of the higher rates obtained at low pressures for a given feed ratio, which in turn is connected with surface coverage, as discussed above. There is also a maximum in productivity at each pressure as the feed ratio is changed. This was to be expected, since there are maxima in reaction rates, although the relationship between those maxima and the ones in productivity is not immediately obvious.

Two crossing lines are shown in Figure 12.7. One follows the line of maximum productivities as total pressure is changed. Clearly, the feed ratio required for maximum productivity increases as total pressure decreases.

At the same time that productivity changes with the feed ratio, the space time (or space velocity) at each point on Figure 12.7 is different. The second crossing line follows the curve of minimum allowable tau values (tau=0.02), as per our chosen constraints. Points to the right of this line are accessible since they lie at or above tau=0.02. This means that the maximum productivity is not accessible at a total pressure of 0.1 bar.

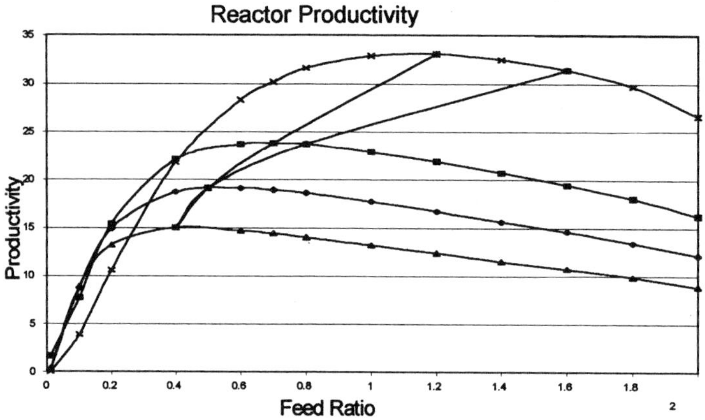

Figure 12.7
Productivity curves at various reaction pressures and feed ratios. The trajectories of maximum productivity and minimum space times are shown by the two crossing curves.

From discussion to this point we see that in order to maximize productivity at the above total pressures, we must operate at the lowest pressure investigated, the maximum allowed temperature, and the maximum allowed space velocity (minimum tau). This is not all there is to the issue of productivity dependence on pressure. Productivity at total pressures below 0.1 bar will behave quite differently, as we will see below.

2. *Maximize the conversion of CO in terms of moles converted per unit of reactor volume while maintaining 99% conversion per pass in a stream at 90% dilution.*

Dilution causes the maximum in productivity to decrease in keeping with the increase in the dilution ratio. The result is that at 90% dilution productivities at the same total pressures as in Figure 12.7 are now those shown in Figure 12.8.

Examination of Figures 12.7 and 12.8 shows that the 1-bar curve on Figure 12.8 corresponds exactly to the 0.1 bar curve in Figure 12.7 except that in Figure 12.8 productivity is about one tenth that in Figure 12.7. This is due to the fact that the amount that can be converted, defined on the total feed rate of reactants, is reduced by a factor of ten by the introduction of the diluent. Curves at pressures other than 1 bar show the effects of decreasing rates of reaction due to dilution. Similarly intertwined curves would be observed at zero dilution on Figure 12.7 at pressures one tenth of those in

Figure 12.8. However, the productivities on the curves in Figure 12.7 would be ten times higher. When it comes to the regions of the curves that are accessible at this dilution within the space velocity limits, things become quite disorganized. For total pressures of 0.5, 1 and 2 bar, tau values of 0.02 and above lie to the right of a feed ratio of roughly 1.5. This means that at the highest allowable space velocities one can operate near maximum productivity at these pressures.

In contrast, at a pressure of 0.1, neither the lower nor the higher space time limit is pertinent and all required space velocities for feed ratios between 0 and 2 are too low (i.e., the space times required are too high) to fall within the stipulated limits of 0.02<tau<0.2 seconds. Conversions of 99% simply cannot be attained with this feed within the range of the stipulated space time constraints.

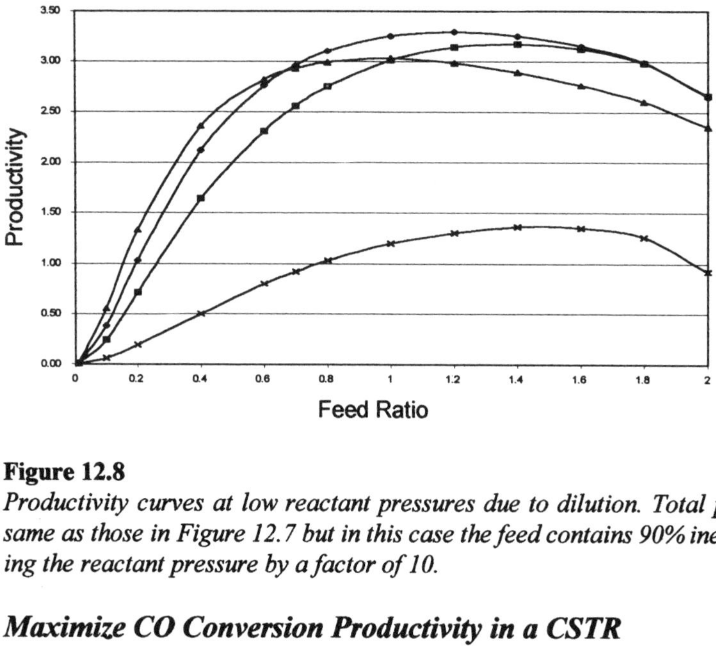

Figure 12.8
Productivity curves at low reactant pressures due to dilution. Total pressures are the same as those in Figure 12.7 but in this case the feed contains 90% inert diluent, lowering the reactant pressure by a factor of 10.

Maximize CO Conversion Productivity in a CSTR

In a CSTR the conversion of the feed is governed by the design equation:

$$-r_A = C_{A0}X_A/\tau \tag{12.7}$$

The term $C_{A0}X_A/\tau$ is the productivity we seek, as defined above. We see at once that the productivity we defined is simply the average rate of conversion in the reactor.

To optimize this productivity while maintaining the required 99% conversion, we seek a CSTR that is operating at the highest rate possible within the constraints and yet produces 99% conversion. It is also clear that we will want to operate at the highest allowable temperature. We have seen for example that in Figure 12.1 there happens to be a condition where maximum rate of about 50 is found at about 99% conversion at 1 bar and a feed ratio of 2. This is already a significant improvement over the highest productivity available in a PFR as shown in Figure 12.7. For a production of 0.666, which takes place at these conditions with zero dilution, the productivity formula requires a space time of 0.013. This is somewhat lower than that allowed under the assumed constraints for PFR operation but, because pressure drop is not a limiting factor in a CSTR, it might well be acceptable in CSTR operation.

A quick search for the highest productivity possible while accepting the temperature and pressure limitations enumerated above results in Figure 12.9.

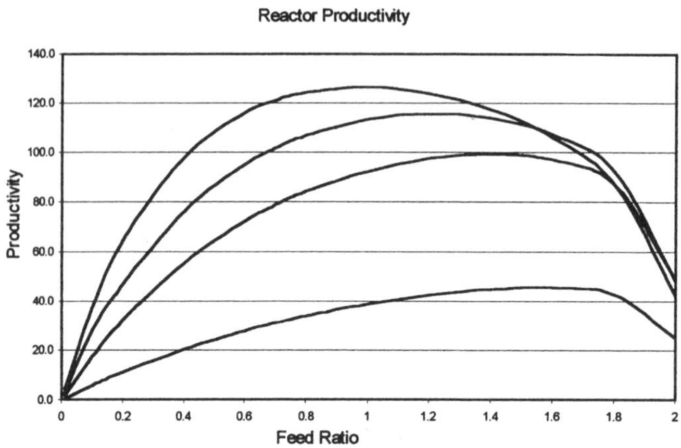

Figure 12.9.
The uppermost curve is for the highest pressure of 2 bar. Lower pressures yield curves lying below that.

The upshot is that an optimized CSTR, operating at 570K and obeying constraints other than that for the space time, will have a productivity four times higher than that achievable at the best of the PFR operating conditions examined previously. The reason is not hard to understand. A CSTR operates at the same maximum rate throughout its volume whereas even the best PFR conditions require an averaging of rates, including ones lower than the maximum that will occur at a single cross section as the feed proceeds through the reactor.

The issue of the space time required for the highest productivity to be achieved in a CSTR may present a design problem despite the fact that pressure drop is insignificant in this type of reactor. At the highest productivity of about 126 in Figure 12.9 the required space time is about 0.0053 seconds. This short space time implies that a large volume of feed is to be passed over a small amount of catalyst. There is a danger that such a combination will result in inadequate mixing of fluids in the reactor, making it an imperfect CSTR, and/or inadequate contacting of gas phase reactants with the solid catalyst, introducing mass transfer inefficiencies. In order to implement realistic operating conditions one may have to back off from the conditions yielding maximum productivity, but even then a CSTR is almost sure to give better productivity than a PFR for this reaction.

The mappings and their examination outlined in the above discussion have led us to a deeper understanding of what is going on in the reaction. Now we are ready to apply this knowledge as we consider ways of improving the kinetic performance of the catalyst.

Designing Catalysts to Improve Performance

The ultimate application of the understanding of reaction rates and their underlying mechanisms lies in the design of catalyst modifications on a rational basis. This is not as novel as one might think, since catalyst modification on the basis of available sources of understanding has guided catalyst development over the years. For example, understanding of the constraints on reaction rates introduced by diffusion has been behind frequent efforts to control the pore size distribution in order to minimize diffusion limitations.

At the same time, the understanding that the rate constant k_r contains a term for the concentration of active sites per unit of catalyst volume or weight has led to efforts to increase the specific surface area of catalysts. The tensions introduced by the desire to increase surface area by increases in porosity and the need for pore size control to avoid diffusion constraints has resulted in a great deal of activity, understanding, and materials development.

What has not been dealt with at nearly the same level of intensity is the relationship between adsorption and catalysis, and in particular between the energy terms in the parameters of the rate expression and catalyst performance. There are good reasons for this neglect, not the least being the lack of a clear and measurable relationship between adsorption and the rate of reaction - the activity - on a given catalyst. This is the problem that mechanistic rate expressions are well suited to illuminate by simulation, although the direct measurement of adsorption isotherms pertinent to the active species on the catalyst surface is not yet possible. At some future time the use of a TS-SSR (see Chapter 5) may give us a better appreciation of the adsorption properties of catalysts.

Assuming that a correct mechanistic interpretation has been achieved in the case of our example of the catalytic oxidation of CO, we now examine what changes in the parameters of that rate might be beneficial, and how one might try to bring about these changes.

The Overall Rate Constant k_r

Theory indicates that the overall rate constant k_r in equation 12.1 contains the square of the active site concentration (as per equation 12.6), the frequency factor for the surface encounters of adsorbed CO with adsorbed O, and the activation energy for the conversion of the two adsorbed species to the gas phase product CO_2.

Altering the number of sites per unit of catalyst

As noted, efforts are frequently made to increase catalyst "activity" (that is, reaction rate at a given test condition) by increasing the number of active sites per unit of catalyst, weight or volume. We will elaborate later on the dangers of misinterpretation consequent on using too simple a measure of activity. In the case of changes in site concentration, however, even standard one-point tests of activity are valid. Often there are reasons to believe that the number of sites per unit surface area is unchangeable. In such cases efforts are concentrated on increasing the surface area per unit of catalyst weight or bulk volume. This leads to issues of pore size distribution, diffusion limitations, catalyst bearing strength, catalyst attrition, sintering and so on. In catalyst development a complex set of criteria pertaining to each of these characteristics has to be optimized within a set of constraints. There are many millions of high-talent man-hours spent in the pursuit of this seemingly straightforward attempt to enhance the activity of catalysts by manipulating catalyst porosity.

Changing the frequency factor

The frequency factor of a catalytic reaction describes the maximum possible number of encounters between the two reactants on the surface under the given conditions (see Chapter 9). In bimolecular reactions this number depends on the density of sites on the surface, since sparse site densities make it unlikely that the two reactants will find themselves on adjacent sites that are close enough for reaction to take place. Thus we can expect that changes in site density can have a disproportionate effect on the rate of reaction via the frequency factor. Doubling site density in a reaction of the kind under discussion may increase the rate of reaction by more than the factor of 4 expected on the basis of site concentration alone, due to the effects of increased density on the entropy of activation and hence the frequency factor of k_r'.

One way it may be possible to increase the frequency factor alone is by inducing the sites to form in "patches" on the total surface. This would make site density beneficially high within the patches without an increase in the total number of sites per unit of catalyst, as measured by some kind of test, such as adsorption. Theoretical studies of adsorption have considered both uniform and patch-wise site distributions in physical adsorption. The consequences of this dichotomy need to be understood as they relate to catalysis.

Rates of monomolecular reactions, those proceeding by the reaction of a single adsorbed species, or of bimolecular reactions between an adsorbed species and a gas phase species proceeding by the Rideal mechanism, will vary linearly with site density and are not expected to depend on the type of site distribution present, homogeneous or patch-wise.

Changing the activation energy of the surface reaction

The activation energy of the surface coupling of adsorbed CO with adsorbed O, as contained in the exponent of the rate constant k_r', is not related to similar couplings that might be studied in the gas phase. It is therefore difficult to foresee what factors may influence its size. Possibly *ab initio* calculations of molecular potential will reveal what changes in catalyst formulation will alter this value.

For now we can only say that a reduction in the activation energy ΔE_r alone, however it may be achieved, will increase activity at a given condition. There is, however, one factor that can be expected to be generally true: any change in the activation energy ΔE_r is likely to be associated with changes in the adsorption isotherms of the active species. Exactly how the two may be related needs to be examined.

It is generally expected that changes in site density affect the rate constant k_r without the alteration of adsorption equilibria. The preconception that changes in catalyst formulation influence only the rate constant k_r gives rise to the use of single point, or standard test, evaluations of catalysts. If it were true, it would cause the kinetic surface to be raised proportionately at all reaction conditions and would allow a single point evaluation to correctly reflect the changes in catalyst performance.

Conversely, changes in any of the energy terms in the rate expression can be expected to result in changes in the energy terms of all the rate parameters. There is no reason to believe that changes in catalyst formulation will not change the energy terms involved. As a result, the shape, not just the level, of the kinetic surface will usually be altered by changes in catalyst formulation, making standard tests at a single point potentially misleading.

The Adsorption Isotherms

The differences in adsorption isotherms as they relate to the distinction between the molecular adsorption of CO and the dissociative adsorption of O_2 to form adsorbed atomic O is well described in standard text books. Here we will accept as fact that in the catalytic oxidation of CO both reacting species are competitively adsorbed on the same sites and that the product CO_2 does not compete for adsorption on these sites.

Isotherms can be separated into two terms, the numerator and the denominator. These can each have a variety of behaviours and need to be considered in turn.

The numerator of the adsorption isotherm

The numerator (see equations 12.4 and 12.5) consists of a term for the gas phase concentration of the precursor of a reacting species and an equilibrium constant, K_a, for the adsorption of the resultant surface species. The first is outside the realm of catalyst design, while the second can be expressed as a product of a pre-exponential and an exponential energy (more correctly, enthalpy) term.

$$
\begin{aligned}
K_a &= \exp(-\Delta G_a / RT) = \exp(\Delta S_a / R - \Delta H_a / RT) \\
&= 10^{13} \exp(\Delta S_{ac} / R) \exp(-\Delta H_a / RT) \\
&= A_a \exp(-\Delta H_a / RT)
\end{aligned}
$$

The pre-exponential A_a is clearly associated with entropy effects defined by the entropy of adsorption, ΔS_{ac}, and may well be driven by the behaviour of the enthalpy term ΔH_a, due to the action of a compensation effect, along the lines discussed in Chapter 9. The term ΔS_{ac} accounts for the entropy of activation and is the entropy term discussed in connection with Figure 9.1. We will return to this point later in this chapter.

The enthalpy in turn depends on the difference between the activation energy of adsorption and desorption. Since the activation energy for adsorption is normally zero while that for desorption is significant, the heat of adsorption is normally exothermic with the result that the exponent of the equilibrium constant is positive and the numerators in isotherm equations 12.4 and 12.5 grow smaller with increased temperature.

As we saw in Chapter 9, the enthalpy of adsorption, or rather of the chemisorption generally involved in catalysis, need not always be negative. It is possible to imagine endothermic adsorption where the numerator of the isotherm is negative and therefore the exponential enthalpy term increases with temperature.

In the case under discussion, both of the adsorption equilibria have negative enthalpies, resulting in positive exponents, with that for the dissociative adsorption of oxygen being ten times larger than that for the molecular adsorption of CO. At the same time the pre-exponential of the oxygen atom equilibrium is seven orders of magnitude smaller, indicating the presence of severe stearic requirements for the dissociative adsorption to take place.

The upshot of these differences is that coverage by O, which is more sensitive to a temperature increase, decreases with temperature, but surprisingly, coverage by CO increases. At low feed ratios the coverage for CO is high while that for O is low. The curves for the two coverages migrate vertically in opposite directions as feed ratio is decreased, until at a ratio of 0.02 they have crossed and most of the surface is covered by O. However, total coverage decreases steadily, as one would expect whenever both reactants are adsorbed exothermically. Figure 12.10 shows the changes in coverage with temperature at 1 bar and a feed ratio of 0.02.

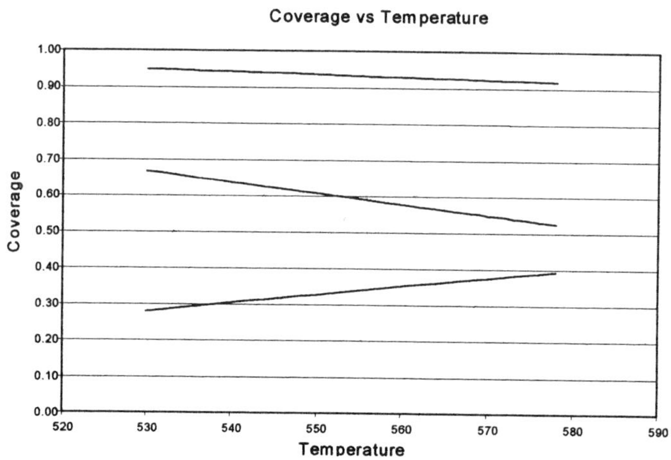

Figure 12.10
The effect of temperature on surface coverage. The upper curve shows total coverage, the next, coverage by O, while the lower curve shows coverage by CO.

From previous discussion we see that the overall rate of reaction, the activity, will be increased by an increase in the numerators of the isotherms or by a decrease in the denominator. Leaving the denominator aside for now, the numerators can be increased by increasing the frequency factors of the isotherms or by increasing the energy term. These changes probably cannot be uncoupled. Previously we had observed that increased exothermicity of adsorption is accompanied by a decreased pre-exponential factor due changes in the entropy of adsorption and vice versa. This is brought about by a compensation effect shown in Figure 9.1 in Chapter 9

Because the correlation line shown in Figure 9.1 goes through the origin, the expected consequence of increasing the enthalpy term is that the configurational entropy of activation will increase by the same factor. If so, then an increase in the exothermicity of adsorption should have an effect on catalyst activity due to two influences.

1. It will change the overall rate at a given temperature due to an decrease in the overall activation energy.

$$-\Delta E_{net} = -\Delta E_r + \Delta H_{CO} + (½)\Delta H_{O2}$$

2. It will change the overall frequency factor

$$A_{net} = A_r\, A_{CO}A_{O2}^{1/2}$$

which implies

$$\Delta S_{net} = \Delta S_r + \Delta S_{CO} + (½)\Delta S_{O2}$$

up or down, depending on whether the configurational entropies of adsorption, ΔS_{Xc}, are positive or negative. For example, an increase in the enthalpy of adsorption can result in a decrease in the frequency factor for adsorption.

In the case of the fit reported in Table 11.1, an increase in the heats of adsorption for the two reactants, assuming there is no change in the parameters of k_r, might result in the parameters of a "new" catalyst shown in Table 12.1. The units used are the same as in Table 11.1 and both the energy terms and the entropy terms pertaining to adsorption have been increased by 1% for the new catalyst.

These changes are fairly arbitrary and need not be defended here. We have already mentioned that there is good reason to believe that the parameters of k_r will change whenever the adsorption characteristics are changed. After all, the bimolecular reaction of adsorbed species most likely depends on the strength of their bonding to the adsorption (active) sites. Moreover, both heats of adsorption will probably not change by the same factor. The two adsorbed species are different and although the chemisorption sites are the same, bonding strengths (and hence heats of adsorption) will surely not show the same sensitivity to changes in the nature of the sites.

Table 12.1
Modifying a Catalyst by Changing Adsorption Characteristics

Parameter	Old Catalyst	New Catalyst
A_r	1.44E+16	1.44E+16
ΔE_r	1.46E+05	1.46E+05
A_{CO}	6.83E+01	5.16E+01
ΔH_{CO}	-7.49E+03	-7.56E+03
A_{O2}	1.99E-06	1.27E-06
ΔH_{O2}	-8.30E+04	-8.38E+04

We will avoid these complications for lack of information on the scale and nature of these effects. The issue we want to consider is: how sensitive is the catalyst likely to be to changes in adsorption characteristics?

One way of examining this is to look at a rate versus conversion plot under the same conditions as those for Figure 12.2. The corresponding figure for the new catalyst is shown in Figure 12.11.

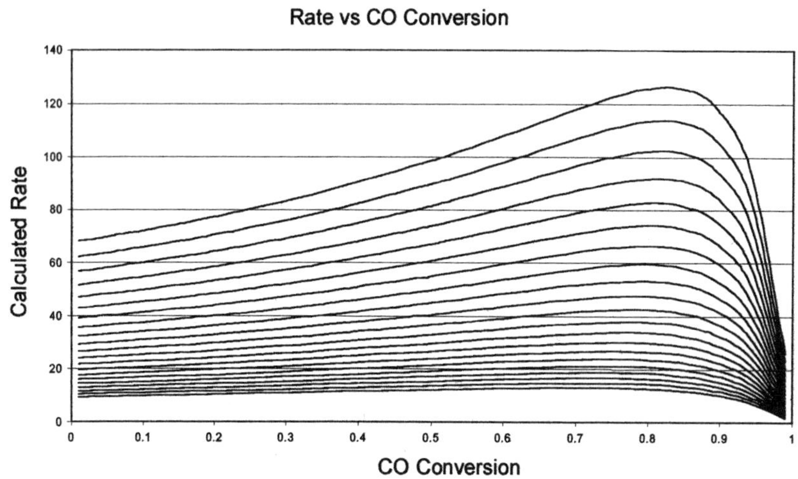

Figure 12.11
Rate versus conversion at conditions corresponding to those for Figure 12.2 but using the parameters for the "new" catalyst in Table 12.1.

Comparison of the two figures shows that the maximum rates and their position change little. In both cases at these conditions we are approaching the maximum possible rate, which is the same on both catalysts since k_r' has not changed. The more important change is in the rates at low conversion, where the reaction is somewhat faster on the new catalyst. Because productivity in a PFR depends on the average reaction rate along the reactor, this difference influences catalyst productivity. Figure 12.12 shows productivity at 1 bar and 570 K for the two catalysts over a range of feed ratios. The maximum productivity has improved by about 5% and is now available at a different feed ratio.

Here it is opportune to illustrate once again the dangers of over-simplified tests in catalyst evaluations. A standard run, if it is carried out at a low feed ratio, will mislead the observer to believe that productivity is much reduced on the new catalyst (see Figure 12.12). The fact is that at higher feed ratios productivity is enhanced. At the same time a standard run done at a high conversion (see Figures 12.11 and 12.2) would show little or no difference in activity between the two catalysts, while one done at low conversion would show a small, perhaps insignificant, improvement on the new catalyst.

The fact is that on the new catalyst activity is increased slightly throughout the range of conversions. In a PFR this increase can result in significant productivity benefits. There is a lesson to be learned here: single point evaluations are often capable of yielding misinformation about catalyst behaviour and always present a potential source of costly confusion in catalyst development.

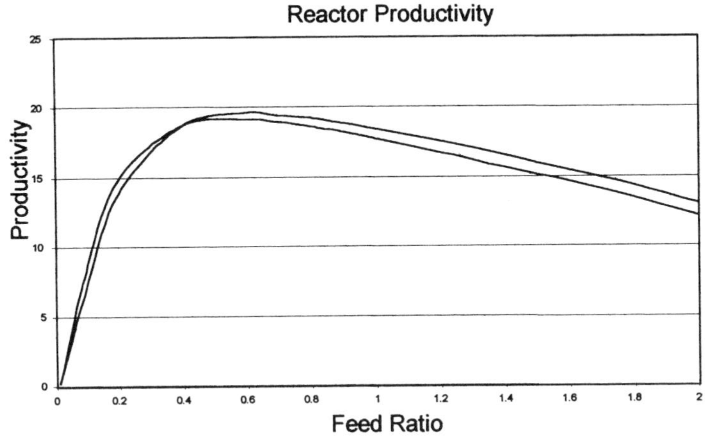

Figure 12.12
Improvement in catalyst productivity due to the lowering of heats of adsorption. Old catalyst productivity is that shown by the lower curve on the right.

There is obviously a large and largely unquantified and unexplored range of kinetic phenomena that can be induced by small changes in the adsorption characteristics of catalysts. A data base relating adsorption equilibria to reaction rates would be needed to make a more informed guess as to how these two aspects of surface activity operate here or in other reactions. Such a data base would be useful to catalysis research under all circumstances, but it would be particularly useful if it could be related to mechanistic rate expressions. Unfortunately, this information is not commonly available, largely because mechanistic rate equations are seldom identified. Nor are the techniques for correlating the information regarding catalyst chemical/physical properties with kinetic rate phenomena well developed at present.

The denominator of the adsorption isotherm.

The denominator of the adsorption isotherms can display a variety of behaviours depending on the sizes of the adsorption terms. The consequent effects on rate behaviour are large if surface coverage is significant and small otherwise. Two limiting conditions can be examined:

- the adsorption equilibrium terms are both much smaller than 1;
- one or both of the equilibrium terms is much larger than 1.

"Much" in this case is a factor of five of more. In the first case, coverage by both species is very small and much of the surface is empty. Notice that a low rate under these conditions is not a problem of inadequate site density but rather of sparse site coverage. From earlier discussion it is clear that rates on such a surface are below their potential maxima. In cases of this sort adsorption should be enhanced and total coverage increased if the material in question is to be improved as a catalyst. Rates will not be increased much by simply increasing the specific surface area, although such a solution will obviously have some beneficial effect. Rather, it will be necessary to change the adsorption equilibria by changing the equilibrium isotherms.

This is easily and commonly done by increasing reaction pressure. However, the required pressures may be costly to implement, encouraging efforts to find a catalyst that allows a higher surface coverage at low pressures. This calls for changing the adsorption properties of the catalyst, a goal that may not be easy to achieve and will require either an increase in the exothermicity (enthalpy) of adsorption or an increase in the frequency factor of adsorption, or both. How to do this is uncertain, but at least if we have clearly identified the problem we know what needs to be done.

The second case is the one that is applicable to the reported results for CO oxidation. To get a sense of the magnitudes of equilibrium parameters involved, the two terms in the denominator for the optimum parameters from Chapter 11, at 1 bar, near zero conversion, at 570 K and a feed ratio of 0.2 are:

$$(K_{CO}P_{CO}) = 54.0 \text{ and } (K_{O2}P_{O2})^{1/2} = 8.11$$

for the old catalyst; and for the new:

$$(K_{CO}P_{CO}) = 42.0 \text{ and } (K_{O2}P_{O2})^{1/2} = 7.11$$

We see again, as we did in Figure 12.9, that the surface is well covered and that adsorbed CO dominates in both catalysts under initial reaction conditions. One can manipulate either term by changing one or both the reactor pressure and/or adsorption equilibrium constants of the catalyst. Note that the disparity between these two numbers on each catalyst is the reason for the low initial rates that would be observed for this feed ratio before a maximum rate is achieved at partial conversion.

Things look different if the feed ratio is reduced. At a feed ratio of 0.03, the "old" catalyst will have the following adsorption parameters:

$$(K_{CO}P_{CO}) = 9.43 \text{ and } (K_{O2}P_{O2})^{1/2} = 8.75$$

Although the equilibrium for CO is now much smaller than at a feed ratio of 0.2, the surface is still highly covered, since both numbers are significantly larger than 1. Moreover, under initial conditions it is now covered almost equally by the two species, since the two equilibria are very similar.

Detailed examination of adsorption behaviour from various perspectives helps to develop an intuitive grasp of the effects of adsorption on this catalyst, and on catalysis in general. Moreover, simulation of a wide variety of kinetic properties in various coordinate mappings should be a high priority in the development of commercial catalysts as well as being the centerpiece of academic research.

Progress in understanding catalytic rates and mechanisms depends greatly on a parallel understanding of adsorption phenomena. It is a pity that the two fields of study have become uncoupled. There is much promise of improved understanding of catalysis in their re-integration.

A View of the Future of Kinetic Studies in Catalyst Development

The kind of thinking that might, even should, follow the determination of a mechanistic rate expression has been presented above. Traditionally, and proudly, the developers of catalysts have perceived their work as an art, albeit laced with a qualitative understanding of catalytic mechanisms based on theoretical work done in the first half of the twentieth century. Much has been accomplished in this way. This text raises the possibility of a major departure from this established view of catalyst development methods. It puts forward a plan to make catalyst development a science.

Several improvements in the last half century in kinetics-related techniques combine to make possible a step change in the study of reaction kinetics, and in particular in the development of catalysts. For example, both analytical and computational methods have changed substantially since the early days when catalytic reaction mechanisms were first formulated. More recently, the development of temperature scanning methods of data gathering in kinetic studies has provided the means for gathering large volumes of raw kinetic information. It is now possible to think of doing rate studies on a scale that is broad enough to allow the determination of believable mechanistic rate expressions for a reaction in a time frame that does not involve a lifelong commitment for an individual or a major expenditure of resources for a research laboratory.

The determination of a mechanistic rate expression by itself will not give all the answers that a catalyst development program may require. Auxiliary data may have to be obtained from studies of adsorption, from established properties reported in handbooks and other literature, or from various surface studies of the adsorbed species. All these may be necessary in an industrial investigation. The mechanistic rate expression by itself will, however, constrain the interpretation of the observations from such auxiliary sources and give significant guidance to the investigation of untested reaction conditions using simulation. It will also allow and encourage the construction of data bases that can guide the rational design of future catalysts.

We have seen how various parameters of a rate expression relate to physical realities in the mechanism of a catalytic reaction. Given this understanding, one can see the benefit of a thorough study of an initial catalyst formulation, one which holds little promise commercially in this early form and would not merit a thorough study using time consuming isothermal methods of rate measurement. The newly developed TSR methods can provide enough kinetic data, in a short enough time, to allow the identification of the correct mechanistic rate expression for the initial catalyst, or at least a rate expression that looks very good under the conditions first tested. A somewhat broader search of the parameter space using the same catalyst will confirm this rate expression or lead to a better, more general form, useable in all future studies involving derivatives of the initial catalyst. All this can be done in days or at most weeks of automated kinetic runs. The most labor intensive, difficult, and perhaps the most time consuming part at this stage of a program is the identification of the correct mechanistic rate expression. Here nothing can replace the intuition, learning and ingenuity of the researcher in charge.

We have argued that, on a homologous series of catalysts, the mechanism of reaction does not change with changes in catalyst formulation. It is therefore worthwhile to spend the time in thoroughly examining an early form of a catalyst and establishing the correct rate expression early on in the program. It will lead to an understanding of the mechanism of the reaction. With this as a guide, one can proceed to change the rate parameters by changing the properties of the catalyst, or reaction conditions. We have seen above how the thinking at this stage of development might go and that the work should finally yield a cohesive data-base for this catalyst.

What we have not seen is how to implement the catalyst modifications necessary to improve some desirable characteristic, such as activity. This is where the process of data-base building comes in. A small series of materials, formulated according to some prior knowledge and using factorial design in a way similar to that used in some high throughput catalyst screening methods, must be investigated using TSR methods. Auxiliary studies must be done as required. From the fitted rate parameters of *the same rate expression* for each of the catalysts thus prepared, changes in the parameters with composition or catalyst treatment are observed. This allows us to form a set of preliminary correlations between catalyst properties and the kinetic parameters.

The above correlations are then used to indicate the direction for subsequent catalyst formulation changes. The results of these experiments are in turn included in correlation building and the procedure iterated as many times as required. Throughout this process it is clear what is being changed, how, and to what end. This is more than one can expect from programs using tests of randomized samples at one standard condition. In the end, an optimum formulation will be found for this catalyst using the procedures outlined above. At that point, on the basis of the understanding gained, one might be prepared to consider what additives or entirely new materials may lead to similar properties while offering improved performance over that of the originally discovered catalyst.

There is still room for art in this style of catalyst development. It will usually come up in the first flush of the discovery of a new catalyst, when major effects are visible from sample to sample. Moreover art or intuition will normally guide the creativity of the researcher in making decisions requiring judgment. However, the refinement and rational design of previously optimized catalyst formulations, those well established in industry, will surely best be done on the basis of good science.

> *The improvement of well established catalysts, where incremental increases in performance are very valuable but very difficult to achieve without a reliable quantitative guide as to how to proceed, is the area where the systematic approach to catalyst development outlined above promises to have the most impact.*

13. TSR Hardware Configurations

General

The essential components of a TS-PFR and of a TS-CSTR are similar to those of the corresponding PFR and CSTR reactors in their traditional configuration. In fact, by applying a lot of manual operation one could, in principle, operate a traditional PFR or CSTR as a TSR. That would be missing a critical advantage of the TSR.

A TSR is ideally suited for thorough integration with computer control combined with data logging and processing. When a continuous-feed reactor (a steady state reactor in its traditional mode of operation) is built as a TSR, with appropriate automation and data processing capabilities, it becomes a *Kinetics Instrument*. Since this conversion from steady state manual operation to a kinetics instrument is the essence of the advantage of a fully functional TSR, we begin by examining the requirements of a TS-PFR and a TS-CSTR in the full TSR configuration.

Figure 13.1 shows a schematic diagram of the components of a steady state reactor. A corresponding TS flow reactor will contain those same items, each under the control of a computer program.

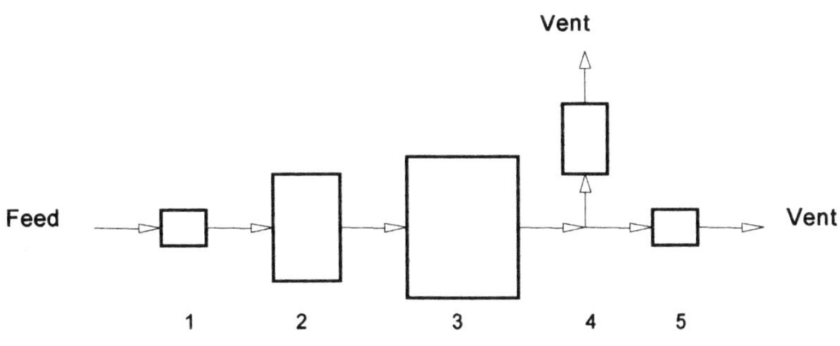

Figure 13.1
Components of TS-PFR and TS-CSTR reactors requiring automation, control and data logging.

The batch reactor (BR) is distinct from the PFR and CSTR, which are steady state reactors, in that it is a transient reactor and in both its conventional and its TSR configuration involves a different operating procedure for gathering data. The temperature scanning stream swept reactor (TS-SSR), mentioned in Chapter 5, also requires a modified procedure since it is both a flow and a batch reactor at the same time. Both the TS-SSR and the TS-BR will be discussed separately from the TS-PFR and TS-CSTR.

The Pre-Heater Module

The critical issues in pre-heater construction are:

1. precise control of temperature of the feed leaving the pre-heater;
2. adequate heating capabilities to maintain the above control at all envisioned flow and ramping rates;
3. small hold-up volume so that reactions in the pre-heater are minimized;
4. good heat transfer to the fluid to minimize temperatures at the heat exchange surface.

The mechanical arrangements will vary with reactor/reaction operating requirements that will dictate pre-heater design, but a useful operating feature in most cases is the capability of a rapid cool-down at the end of each run. To this end a small thermal mass is desirable. The optimum design might therefore involve delivery of heat by preheated air to a thin tube through which the reactant is flowing. To avoid large energy demands and hot air venting problems this can be done inside a recirculating-air oven, fitted with suitably powerful heaters, and containing a coil of tubing to transport the feed. Such a design has the advantage of having a small thermal mass and, given adequate heaters and other design features, a rapid response to control inputs. Besides offering a rapid cool-down, such an oven is also able to follow ramping profiles precisely. This should satisfy issues 1 and 2 listed above.

If the coil of pre-heater tubing is of adequate length, the temperature inside the pre-heater, and therefore the wall temperature inside the coil, need not be much higher than the required outlet temperature. This should satisfy issue 4 listed above.

In order to minimize the dead volume in the pre-heater, small diameter tubing should be used. This will not only minimize wall thickness for a given operating pressure but will also facilitate heat transfer by increasing the turbulence in the feed and reducing heat transfer resistance at and in the tubing wall. These advantages must be balanced against the increased pressure drop in a small-ID pre-heater, a problem that can lead to several complications. For example it may impose unacceptable pressure requirements at the metering devices, or lead to unreasonable wall thickness requirements in the pre-heater coil, or make it hard to define the actual reactor pressure.

In a versatile TSR the pre-heater should be readily replaceable as the demands change. Furthermore, the changes must be such that the existing heaters can heat the reconfigured pre-heater. An air oven containing a pre-heater coil is again a suitable answer though a fluidized-solid or a liquid bath may be appropriate in some cases.

Traditional heating coils, embedded in ceramic clam shells, are generally too slow to react to the changing temperatures required in TS-PFR operation. Such a configuration may, however, be cheaper and may serve well in the case of a TSR where precise input temperature ramping is not critical, or where high temperatures are required, or if the ramping rates to be used are low. Clam shell furnaces will usually be much slower in cool-down mode and may be more useful in up-down ramping operations of the TS-PFR described in Chapter 5.

The Reactor Module

The reactor module will require a heater whose configuration is dependent on the kind of ramping to be used and on the shape of reactor body to be heated. We first consider a PFR configured as a straight tube of constant diameter. Other configurations, such as coils and non circular cross-sections or discs (very short tubes) are rare and their construction can also be guided by the following ideas. It is the modes of operation that will have the greatest influence on the design of this module. In Chapter 5 we considered several modes of temperature ramping in TS-PFR operation. We also considered some modes of operation involving flow-rate changes. Only the temperature ramping options weigh heavily on the design of the reactor module. These are:

1. a series of up rampings to a common high temperature during which data is collected, each ramping is followed by an idle cool-down period to a common, low, starting temperature;
2. a series of up rampings to a common high temperature, each followed by an immediate down ramping to a common low temperature after a change in feed rate. Data is collected during both the up and down ramps;
3. a series of up rampings to a common high temperature during which data is collected. Each up ramp is terminated at the same high temperature. This condition is maintained at a new flow rate until steady state is achieved. No data need be recorded during this stabilization time. Next, there is a down ramp from the high temperature down to the common low temperature. Data is collected during this down ramp.

The reactor module for TS-PFRs in general requires a spacious volume so that the reactor body itself can be changed in length and diameter. This need is best accommodated if the volume containing the reactor is larger than that of any of the reactor sizes to be used. This in turn requires a heat transfer medium that will couple the output of the heaters to the reactor regardless of size or configuration.

Media that can serve this need include liquids, gases and fluidized solids. The specific choice will depend in the temperatures expected during reaction and other issues. A significant influence on this choice will come from the heat capacity of the med-ium and hence the difficulty of ramping temperatures rapidly enough for the planned experiments. Gases are therefore to be preferred. Because if the low heat capacities of gases, rapid and efficient circulation between the heaters and the reactor must be assured.

Air ovens and air fluidized beds offer the best opportunities for rapid cool-down between rampings. Liquid baths can be cooled rapidly if cooling coils can be installed in the heated volume or if the same medium can be passed through the coils for both heating and cooling. In either case liquid baths will be restricted to slower ramping rates than air baths.

The theory of temperature scanning allows us to ramp the inlet temperature to a TS-PFR independently of the temperature of the reactor, as long as certain conditions are observed (see Chapter 5). However, there seems to be no special advantage to following this type of operation. The optimum solution is then to ramp both the input feed and the reactor itself along the same trajectory.

This can readily be done if the reactor heater volume also contains the pre-heater. In this configuration not only do we avoid a separate control function for each of the two modules but we improve experimental reproducibility and minimize random errors. To achieve this, the reactor and pre-heater have to be so nested that the output temperature of the pre-heater is the same as the inlet temperature of the reactor. Careful design, appropriate temperature controls and control of heater-fluid circulation in the heater bath can be made to achieve the desired condition. In the design of TS-PFRs this is arguably the most complicated aspect.

Unless there is compelling reason to do otherwise, the optimum design for type 1 operation is therefore an air oven containing both the pre-heater coil and the reactor body. Heaters and circulation must be properly configured and a blast of cold air made available for rapid cool-down between rampings. To enhance both the heating and cooling rates the interior components of the oven should be designed with a view to minimizing the thermal mass in the hot zone.

In order to accommodate a variety of reactions and reaction conditions, both the reactor body and the pre-heater should be readily accessible and dismountable so they can be replaced. This calls for the use of high temperature connecting fittings between the reactor and pre-heater to the externals of the oven. One variant of this design involves a back plate in the oven that can be withdrawn from the oven itself making all internal fittings connecting the reactor, pre-heater, temperature sensors and sampling ports readily accessible for alterations.

In the case of up-and-down ramping, as described in type 2 operation, the reactor setup is very similar to that described above for type 1 operation. The only substantive difference is that the cooling part of each cycle must be under strict temperature control so that the cooling temperature trajectory is the same after each up-ramp. This involves an adequate supply of well distributed cooling medium to deal with the thermal mass of the unit at projected cooling rates. It also requires coordinated control of cooling and heating during the down-ramp, so that the desired temperature trajectory shape can be maintained.

Type 3 operation, involving ramping up followed by a stabilization at the high temperature and then a cooling down to the lower temperature, simply increases demands on the cooling and control capabilities described for type 2 operation. The additional high temperature soaking during the stabilization period increases the heat content of the thermal mass in the oven and calls for more cooling capacity and even better coolant distribution than type 2 operation.

The reactor module will normally contain several controlled elements as well as several temperature measuring devices. All of these must be capable of communicating with the TSR software. The controlled elements should accept commands and report status to the TSR software. The temperature measuring devices will only report.

The pre-heater/reactor module can be so constructed that the reactor and preheater inside the unit can be readily replaced as the direction of research interests changes, without disturbing the feed and analytical sections. The versatility this will give a given setup is worth the effort in a laboratory that pursues varied programs, for example at a university.

The Analytical Module

The speed of data acquisition using temperature scanning methods is limited by the speed of sample analysis. As a rough rule of thumb it is advisable to take one sample per degree of temperature change in the range of temperatures to be scanned. For a typical temperature range of 100 C° one therefore needs 100 analyses per ramp. For a typical experiment of 10 ramps 1000 analyses are required. If analysis time is 6 seconds per sample, one obtains 10 analyses per minute and all ten ramps can be completed in 100 minutes. Add to that 10 minutes per ramp for cool-down and stabilization to arrive at approximately 200 minutes per experiment, about 3½ hours. This allows for six experiments, perhaps using different feed compositions or pressures, in a single 24-hour period of automated operation. Such a volume of data represents a very thorough evaluation of the kinetics of a system and, after the procedures described in Chapters 5 and 7 have been applied, a thorough understanding of its behaviour. Calling a TSR a *Kinetics Instrument* makes sense in this connotation, given that the data is made available in semi-processed form in real-time.

Analytical methods vary greatly in speed of data acquisition and a premium should be placed on speed when designing a TSR. For example, in liquid systems at low temperatures, electrodes may provide an adequate means of analysis. Such truly continuous monitoring methods would, in principle, allow very high ramping rates. These, however, will be constrained to more realistic values by the available speed of temperature ramping and the necessity to make sure that everything, reactor, reactant, electrodes, and so on, are at the same uniform temperature at each measurement. The ramping rates of reactors equipped with such truly continuous analytical methods will be restricted by these considerations to values not much different from those using "semi continuous" techniques.

Recently more and more of the traditional methods of analysis are being made in quick turn-around versions, yielding semi-continuous analytical data. This is done using automation and clever mathematical processing of raw data. One such method is the deconvolution of mass spectra of mixed streams. Using this technique, a quadrupole mass spectrometer, and a fairly simple calibration procedure described in Chapter 5, it is possible to quantify as many as ten components in a combined effluent stream at the output of the reactor. The method is presently capable of one analysis every six seconds or so using relatively inexpensive quadrupole mass spectrometers. We use this capability as the basis of the above estimate of time required per experiment. More elaborate and more expensive mass spectrometers will allow the deconvolution of more components while other deconvolution methods, such as deconvolution of IR spectra, are being brought to fruition. These near-continuous methods are recommended for TSRs whenever they can be used.

Slower than this are the speeded-up versions of traditional analytical methods such a liquid or gas chromatography. The length of time of an analysis can vary greatly in these cases, but let us consider a ten minute gas-chromatographic analysis as a base case. This makes the ramping time in the six-experiment investigation envisaged in the lead paragraph 100 times as long, but presumably the cool-down periods would remain at 10 minutes per ramp. The total time of the study would therefore be roughly 11,000 minutes per experiment or 66,000 minutes for the six-experiment investigation. This works out to about 46 days, a month and a half of continuous operation. One might be inclined to cut corners by reducing the frequency of analysis to speed things up, but such impatience is surely unseemly in view of how long a similar investigation would take using conventional steady state methods to cover the same range of conditions.

In all the cases it is important that the analysis be correctly assigned to the conditions prevailing at the outlet of the reactor at the time of sampling. The theory of temperature scanning requires that compatible triplets of temperature-conversion-space time be used for the interpretation of data. This means that samples must be taken right at the outlet of the reactor, rather than some distance away, as is normal in steady state reactors. This in turn may call for ingenuity and novel sampling methods as the sampling device is introduced into the hot and cramped region at the outlet of the reactor.

For example, a capillary sampling tube can be inserted into the effluent at the end of the reactor and led directly to the vacuum chamber of the ionizer of a mass spectrometer. Issues concerning capillary size to avoid overloading the vacuum pumps of the mass spectrometer, of variation in sample flow rate with temperature and composition changes during reaction, of condensation during transit to the mass spectrometer, and other such factors may be important and have to be examined and dealt with to avoid distortion in the results. Random distortion from various sources can be adequately dealt with as noise, in ways described in Chapter 7. Systematic distortion, which can be caused by neglecting any of the above issues, is more treacherous and must be carefully avoided.

If there is any reason to doubt the validity of the sampling procedure, it is best to carry out a thorough investigation of the suspected configuration, examine modifications of the existing configuration, and/or make changes in sampling probe placement, or do all of the above before distorted data is collected and used. The TSR, badly constructed, can be a deceptive instrument, like any other.

The analytical equipment will most often come with its own data processing software and hardware. The output presented at the end of each analysis is therefore in the form of a file reporting, say, mole fractions of components in the sample. In order to carry out an analysis the TSR software must trigger the taking of a sample at appropriate times, simultaneously record the various parameters of the system at those times, and wait for the analysis to be completed on the analytical setup. Then the TSR software must either trigger the transmission of the analysis report or be prepared to accept the analytical report at the time it is automatically sent out by the analytical equipment. In both cases, the analytical file must be assigned to the conditions at the time of sampling.

The capture and subsequent formatting and processing of the analytical report is done under TSR software control. Displays of the results of the analysis and any other pertinent parameters and presentations is under TSR software control from there on. Real-time strip chart presentations and their recordings can be used to observe or review any anomalies that might take place during hours of unattended operation.

The Outlet Controls

Reactor pressure is maintained by a back-pressure controller operating well away from the harsh environment of the reactor module. The only requirement here is that the back pressure controller be capable of responding to set points sent from the multiplexer and of reporting the actual pressure maintained to the TSR software.

Differences Between the TS-PFR and the TS-CSTR

The required modules for both the TS-PFR and the TS-CSTR are identical in function if not in configuration. The major differences are introduced if the TS-CSTR is to be operated in the "joystick" mode, as described in Chapter 5. In that case the control of feed flows must be done using a device offering instant response to command inputs. This is generally not possible with standard flow-meters and additional thought must be given to the selection of flow-control devices.

However, since the response of the temperature control heaters is also constrained to the slow side, it may well be possible to vary flow continuously at a slow but adequate rate using available mass flow meters. Both these issues result in requirements that end up reducing the possibility of instantaneous response to joystick commands and make it necessary for the software to simply record the trajectory commanded and then implement it at rates compatible with the capabilities of the hardware.

The other option is to drive the flow/temperature trajectory as hard as one can using the joystick, and then simply collect data as it comes. Since there is no need for the TS-CSTR to be at commanded conditions when data is collected, it is possible to gather data at any conditions actually achieved during this exercise, and to interpret it correctly, albeit at conditions one did not exactly foresee. All that is required is that the actual conditions be recorded and that a valid CSTR operation is maintained.

The Transient Reactors

The TS-BR

The TS-BR does not require modules 1, 2 or 5 shown in Figure 13.1. The reactor is filled off-line and introduced into the heater assembly. Appropriate probes are inserted to monitor temperature and composition and the ramping proceeds at any desired rate, in fact along any random or desired trajectory. The requirement of TS-BR operation is simply that the several runs constituting an experiment use different temperature trajectories.

To achieve the variety of temperature trajectories required for mechanistic studies and for rate mapping, the reactor contents must be very well mixed, the reactor itself must have high heat transfer capabilities, and it must be surrounded by high output heaters/coolers. The option of linear ramping at several ramping rates is the most obvious protocol to use in most cases, in which case a range of ten or more in ramping rate should be available. The greater the range of ramping rates available, the more detailed the picture of rates made available by the TS-BR.

Since there is no requirement for ramping regularity, data can be collected at all times from ambient temperature up to a specified maximum temperature. This makes available data from zero conversion at ambient temperatures up to some agreed maximum temperature and/or conversion. There is no loss of data corresponding to the heat-up period normally encountered in batch reactor operation.

The duration of each ramp will be controlled by the maximum temperature allowed, or by a specified conversion limit. Since there is no need to collect data after reaction goes to completion or after equilibrium has been established, the duration of useful ramping at a the various temperature ramping rates may be different. This will require placing the ramp duration at each rate under the control of the output from the analytical equipment, as well as limiting ramp duration by a temperature limit.

It is also clear that different sampling rates will have to be implemented at the several ramping rates in order to acquire the recommended one analysis per degree of ramp. In cases where ramping rates are varied by a factor of ten, rates of analysis may have to vary by a factor of more than ten. The limiting factor in a TS-BR experiment will be the highest rate of analysis possible, which in turn will limit the highest ramping rate allowed. To follow the example given in the case of the TS-PFR, if a 100 C° span is to be used, and the highest ramping rate allowed is 10 C°/minute, the fastest ramp will take 10 minutes. Subsequent rampings will take longer until the last ramp, which will take 100 minutes. Cooling and reloading might take about 30 minutes per run. A ten-run experiment can therefore take about 10 hours and would require periodic supervision to reload the reactor. A six-experiment study will take a week of regularly supervised ramping. Automation of the reloading operation would therefore be very useful in enhancing the utility of a TS-BR.

The TS-SSR

The TS-SSR is a temperature scanning reactor intended for kinetic studies of interactions between fluids and solids. It is therefore appropriate for studies of adsorption/desorption, ore roasting, solid oxidations/reductions and so on. Because a TSR reports rates of reaction regardless of the mechanism, a TS-SSR will report reaction rates in all diffusion regimes. Mechanistic studies seeking to determine the rates of interaction between the solid surface and the fluid free of diffusion limitations will have to be proceeded by careful delineation of conditions that allow diffusion-free measurements to be made.

The TS-SST is a reactor combining the characteristics of a flow reactor with those of a transient reactor such as a BR. This alerts us to the fact that the design and hardware requirements of a TS-SSR have to take into account all the features and operational requirements of both the PFR/CSTR type and BR type of reactor. It also suggests that the operating protocol of a TS-SSR can be like that of a BR or of a flow reactor.

There are two kinds of TS-SSR: the plug flow version which we have called the TS-PF-SSR, and the back-mix version which we call the TS-CST-SSR. The modes of operation, theoretical underpinnings and data processing methods appropriate for each of these two SSR types are described in Chapter 5. The hardware is described below.

Operating the TS-PF-SSR

The TS-PF-SSR is essentially identical to a TS-PFR designed for the study of heterogeneous catalysis. The differences are in the details of design of the reactor body itself and in the mode of operation. Whereas the kinetic TS-PFR experiment consists of several runs at the same ramping rate and several different feed rates, the TS-PF-SSR adsorption experiment can consist of several runs at different feed rates and the same ramping rate, or at several different ramping rates and the same feed rate.

In order to study pure rates of desorption, temperature scanning rates in a TS-PF-SSR should be coordinated with flow rates so that a predetermined maximum change in concentration in the fluid phase is not exceeded. This is necessary to limit the extent of back-reaction and thereby determine the rate of the desorption reaction in isolation. The preferred procedure to do this is to vary flow rate during a constant-rate temperature ramping so that the sweeping stream does not exceed a preset concentration. Interpretation of data is based on the amount of material released/consumed by the solid per unit of time during the ramp (see Chapter 5).

Since effluent analysis will report the concentration of the component of interest in the effluent, this number is used to adjust the flow rate to stay within the preset maximum. At the same time the *total* amount of the component of interest leaving the outlet depends on the flow rate. This amount has to be calculated as the run progresses taking into account the flow rate of the sweeping gas and the concentration of desorbed material at each instant of time.

To reduce axial coverage profiles in the case of the fixed bed TS-PF-SSR used for studying desorption alone, the reactor tube might also be made as short as possible. As mentioned in Chapter 5 this brings up the problem of maintaining an even carrier gas flow through a thin wide bed.

A test for the applicability of the simple interpretation given by equation 5.50a consists of varying the feed rate until one finds feed rates that give the same r_n regardless of τ at all temperatures during the ramp.

In cases where it is possible to chose operating conditions that allow the use of a TS-PF-SSR, the simplicity of that setup has much to recommend it. The TS-PF-SSR not only yields data that is simpler to interpret, but involves a simpler hardware design.

Operating the TS-CST-SSR

The attractions of the TS-PF-SSR configuration notwithstanding, the ideal configuration for a TS-SSR is probably a CSTR operated as a TS-CST-SSR. The advantages of the CSTR configuration have been mentioned previously in regards to kinetic studies in general. In gas-solid reactions these advantages become very clear in at least one case we will consider. However, in adsorption studies, this configuration requires a more complicated data interpretation method than that used for the interpretation of TS-PF-SSR data.

In return for this complication, the TS-CST-SSR can yield very interesting results with much less effort than the TS-PF-SSR. This comes from the very different contacting pattern inherent in the TS-CST-SSR configuration as compared to the TS-PF-SSR, that allows the direct measurement of several quantities of interest. In Chapter 5 we have elaborated on the necessary theory and a mode of operation for the TS-CST-SSR. There it was shown how, in the special case of Langmuirian adsorption, this type of reactor will yield, in one run, more information than is available from a TS-PF-SSR experiment consisting of several runs.

The mode of operation of a TS-CST-SSR in normal circumstances, where the kinetics is unknown *a priori,* is the standard one of doing several runs, each at a different ramping rate and the same sweeping gas flow rate or each at a different flow rate and the same ramping rate. The procedure more or less follows that for the TS-PF-SSR. A review of TS-CSTR operations and those of a TS-PF-SSR described in Chapter 5 will explain what is to be done. However, in the special and interesting case of monolayer adsorption in a Langmuirian adsorption system there is a special treat.

In Langmuirian systems the TS-CST-SSR makes available quantitative information concerning the adsorption/desorption rate constants, the equilibrium constant, all their Arrhenius parameters, the *effective* volume of the sample and the *effective* void volume of the reactor vessel, from a single ramping. To do this one needs to be able to determine the solid's full monolayer capacity in preliminary auxiliary runs and to supply a sample covered by a near-monolayer of adsorbent.

Considering the less-than-quantitative information being obtained from TPD apparata using current modes of operation, one could call such a run the "silver bullet" experiment. Surprisingly, the procedure required to perform the silver bullet experiment is almost identical to that of a TPD run. The difference comes from using a CSTR to contain the solid sample and an entirely new method of data interpretation.

Unlike the TS-PFR and the TS-CSTR, the TS-SSR normally requires that a broad range of temperature ramping rates be possible. In this the two TS-SSRs are similar to the TS-BR and all the issues regarding the speed and frequency of analysis described in the section on the TS-BR apply here too. The option of varying feed flow rate in a series of runs at constant ramping rate, as is done in a standard TSR experiment, is also available if ramping rates can not be varied adequately.

The one exception to this requirement is the "silver bullet" experiment. There, a single ramping can characterize a solid catalyst sample in great detail and make the parameters listed above available for meaningful integration with mechanistic kinetic parameters obtained on the same catalyst sample using a catalytic TS-PFR or TS-CSTR.

The interplay between the results from a catalytic TSR and those obtained from adsorption studies done using a TS-SSR offers a means of connecting adsorption kinetics, and thermodynamics, with catalytic activity and selectivity. Adsorption and reaction on a given catalyst could then be understood in terms of the elementary chemical processes taking place between gas phase species and the surface. Such a view should significantly augment and constrain our understanding of reaction mechanisms as they are currently visualized on the basis of observations of surface species under high vacuum conditions.

> *It is interesting to note that an appropriately equipped TSR can be operated as a TS-SSR. This flexibility would allow adsorption studies of catalyst samples to be made routinely as part of the study of catalytic reaction kinetics.*

Summary

The details of the required design features in the case of the various TSRs described above will be defined in due course on the basis of laboratory experience. The points raised in the above text are intended as a presentation of our current thinking regarding the issues that need to be considered in building an effective TSR. When commercial versions of TSRs are purchased, the user will be spared the need to understand the appropriate mathematics and the underlying theoretical issues. More invitingly, commercial TSRs will spare the user the need to write the programs used to control the operation of the equipment and to interpret the results. The TSR will be an instrument, like any other. Like any other instrument it will give good results, rapidly and painlessly, when operated properly. The real sophistication and complexity of the TSR will be largely hidden from the user, as it is in most instruments.

Given that the user is basically interested in kinetic results, not in the construction and programming of an apparatus, the choice of a manufactured TSR makes sense. An off-the-shelf TSR Kinetics Instrument should require a minimum of esoteric understanding to be operated effectively. Building a home-made TSR will then be as attractive as building a home-made gas chromatograph.

The TSR is an instrument for measuring quantities that have previously been hard-won, at the cost of much time and labor. It is another step in the line of instruments that have replaced labor-intensive data accumulation methods, such as the aforementioned gas-chromatograph. In each such case, when a new instrumental method is introduced, the new technique speeds up progress in research and understanding in one or several fields of work. One can expect that the same will be true with the advent of the TSR.

This book presents the theory and operation of a variety of temperature scanning reactors. To the average user this equipment will one day be just another "black box". The details of the TSR methods and their theoretical justification will fade into the background. What will remain is an instrument that enhances the user's ability to work in chemical kinetics and mechanistic studies.

References

Amenomiya, Y. and R.J. Cvetanovic, 1963, J. Phys. Chem., 67, 144.

Asprey, S.P., 1997, Theory and Application of the Temperature-Scanning Plug-Flow Reactor, PhD. Thesis, Queen's University, Kingston, Ontario, Canada.

Asprey, S.P., N.M. Rice, and B.W. Wojciechowski, 1997, The Temperature Scanning Reactor III: Experimental Procedures and Data Processing, Catalysis Today, 36, pp. 209-226.

Asprey, S.P., B.W. Wojciechowski, N.M. Rice and A. Dorcas, 1996, Applications of Temperature Scanning on Kinetic Investigations: The Hydrolysis of Acetic Anhydride, Chemical Engineering Science, 51, pp. 4681-4692

Asprey, S.P., B.W. Wojciechowski and B.A. Peppley, 1999, Kinetic Studies Using Temperature Scanning: The Steam reforming of Methanol, Applied Catalysis A. General 179, pp. 51-70

Bates, D.M. and D.G. Watts, 1988, Nonlinear Regression Analysis and its Applications, John Wiley and Sons, Toronto.

Bennett, C.O., 1976, The Transient Method and Elementary Steps in Heterogeneous Catalysis, Cat. Rev. – Sci. Eng., 13, 121.

Bisio, A. and R.L Kabel, 1985, Scaleup of Chemical Processes, Wiley, New York, pp. 136-138.

Bonanzi, S., 1996, Smoothing by Basis Function Expansions: A Comparison with Digital Filtering, Master's Thesis, Queen's University, Kingston, Ontario, Canada.

Broyden, C.G., 1970, The Convergence of a Class of Double Rank Minimization Algorithms: Parts I and II, J. Inst. Maths. Applics. 6, pp. 76-90 and 222-231.

Boudart, M., D.E. Mears and M.A. Vannice., 1967, Ind.Chim.Belge, Special Issue, 32, 281.

Candruff, J., 1978, An Introduction to Organic Chemistry, p.167, Wiley, New York.

Cox, M.G., 1972, The numerical evaluation of B-splines, J. Inst. Math. Applics., 10, pp. 134-149.

Cleland, F.A. and R.H. Wilhelm, 1956, Diffusion and Reaction in Viscous-Flow Tubular Reactor, A.I.Ch.E.J, 2, pp. 489-497.

Davidon, W.C., 1959, Variable Metric Method for Minimization, Research and Development Report ANL-5990 (Ref.) U.S. Atomic Energy Commission, Argonne National Laboratories.

Davis, P.J., 1973, Statistics and Data Analysis in Geology, Wiley, New York.

Davis, P.J., 1975, Interpolation and Approximation, Dover Publications, New York.

de Boor, C., 1972, On calculating with B-splines, J. Approx. Theory, 6, pp. 50-62.

de Boor, C., 1978, A Practical Guide to Splines, Springer Verlag, New York.

Dierckx, P., 1982, A Fast Algorithm for Smoothing Data on a Rectangular Grid while Using Spline Functions, SIAM J. Numer. Anal. 19, pp. 1286–1304.

Domke, S.B., R.F. Pogue, F.J.R. Van Neer, C.M. Smith and B.W. Wojciechowski, 2001, Investigation of Ethylbenzene Pyrolysis Using a Temperature-Scanning Reactor, Ind. Eng. Chem. (Kinetics, Catalysis, and Reaction Engineering), 40, pp. 5878-5884.

Draper, N.R. and H. Smith, 1981, Applied Linear Regression (second edition), Wiley, New York.

Eldridge, J.W. and E.L. Piret, 1950, Chem. Eng. Prog., 47, 363.

Fletcher, R., 1970, A New Approach to Variable Metric Algorithms, Computer J. 13, pp. 317-322.

Fletcher, R., 1987, Practical Methods of Optimization, Wiley, Chichester.

Fletcher, R. and M.J.D. Powell, 1963, A Rapidly Convergent Descent Method for Minimization, Computer J., 6, pp. 163-168.

Fletcher, R. and C.M. Reeves, 1964, Function Minimization by Conjugate Gradients, Computer J., 7, pp. 149-154

Glasser, D. and D.F. Williams, 1971, The Study of Liquid Phase Kinetics Using Temperature as a Measured Variable, Ind. Eng. Chem. Fundamentals, 10(3), 516.

Gleaves, J.T., J.R. Ebner and T.C. Kuechler, 1988, Temporal Analysis of Products (TAP) – A Unique Catalyst Evaluation System with Submillisecond Time Resolution, Cat. Rev. – Sci. Eng., 30(1) 49.

Goldfarb, D., 1970, A Family of Variable Metric Methods Derived by Variational Means, Maths. Comput., 24, pp. 23-26.

Grenier, B., 1997, A Treatment of Catalyst Decay Using a Temperature-Scanning Plug Flow Reactor, M.Sc. Thesis, Queen's University.

Hayes, G.W. and J. Halliday, 1974, The Least-Squares Fitting of Cubic Spline Surfaces to General Data Sets, J. Inst. Math. Applics., 14, pp. 89-103.

Kemp, R.R.D. and B.W. Wojciechowski, 1974, The Kinetics of Mixed Feed Reactions, I.&E.C. (Fundamentals) 13, pp. 332-337.

King, R.P. and D. Glasser, 1965, The Use Of Adiabatic Calorimeter for Reaction Rate Studies, So. Afr. Ind. Chem., 19, pp. 12-15.

Lawson, C.L. and R.J. Hanson, 1974, Solving Least Squares Problems, Prentice Hall, Englewood Cliffs.

Lay, D.C., 1993, Linear Algebra and its Applications, Addison-Wesley, Reading Massachusetts.

Levenberg, K., 1944, A Method for the Solution of Certain Nonlinear Problems in Least Squares, Quart. Appl. Math., 2, pp. 164-168.

Levenspiel, O., 1962, Chemical Reaction Engineering, Wiley, New York.

Luenberger, D.G., 1984, Linear and Nonlinear Programming, 2nd Edition, Addison-Wesley, Reading, Massachusetts.

March, J., 1985, Advanced Organic Chemistry: Reactions, Mechanisms and Structure, p. 333, Wiley, New York.

Marquardt, D.W., 1963, An algorithm for Least Squares Estimation of Nonlinear Parameters, SIAM J., 11, pp. 431-441.

Nelder, J.A., and Mead, R., 1965, A Simplex Method for Function Minimization, Computer Journal, vol. 7, 308.

NEOS Guide to Optimization Software, http://www.fp.mcs.anl.gov/otc/Guide/Software Guide/index.html.

Peppley, B.A., J.C. Amphlett, L.M. Kearns, and R.F. Mann, 1999, Methanol Steam Reforming on $Cu/ZnO/Al_2O_3$. Part 1: The reaction network, Applied Catalysis, A 180, pp. 21-29.

Polak, E., 1971, Computational Methods in Optimization: A Unified Approach, Academic Press, New York.

Polak, E. and Ribiere, G., 1969, Note sur la Convergence de Methods de Directions Conjugres, Revue Francaise Informat. Recherche Operationnelle 16, pp. 35-43.

Press, W.H., B.P. Flaherty, S.A. Teukolsky, W.T. Vetterling, 1986, Numerical Recipes, the Art of Scientific Computing, Cambridge University Press, Cambridge.

Rice, N.M. and B.W. Wojciechowski, 1997, The Temperature Scanning Reactor II: Theory of Operation, Catalysis Today, 36, pp. 191–207.

Shanno, D.F., 1970, Conditioning of Quasi-Newton Methods for Function Minimization, Maths. Comput., 24, pp. 647-656.

Shatynski, J.J. and D. Hanesian, 1993, Adiabatic Kinetic Studies of the Cytidine/Acetic Anhydride Reaction by Utilizing Temperature versus Time Data, Ind. Eng. Chem. Research, 32, pp. 594-599.

Watts, D.G., 1994, Estimating Rate Parameters in Nonlinear rate Equations, Can. J. Chem. Eng., 72, 701.

Wojciechowski, B.W., 1997, The Temperature Scanning Reactor I: Reactor Types and Modes of Operation, Catalysis Today, 36, pp. 191–207.

Wojciechowski, B.W., 1999/2000, The Application of Temperature Scanning to Adsorption Studies, Annales, Sec AA Chemia, UMCS, Lublin.

Wojciechowski, B.W. and S.P. Asprey, 2000, Kinetic Studies Using Temperature Scanning: The Oxidation of Carbon Monoxide, Applied Catalysis A. General 190, pp. 1-24.

Wojciechowski, B.W. and N.M. Rice, 1993, Temperature Scanning Plug Flow Reactor, Chemical Engineering Science, 48, pp. 1100-1104.

Index